More Than
Houses

More Than
Houses

*How Habitat for Humanity
Is Transforming Lives and Neighborhoods*

Millard Fuller

WORD PUBLISHING
NASHVILLE
A Thomas Nelson Company

Word Publishing, Nashville, Tennessee

Scripture quotations marked NIV are taken from *The Holy Bible, New International Version*®. Copyright © 1973, 1978, 1984 by International Bible Society. Used by permission of Zondervan Publishing House. All rights reserved. Those marked TEV are from *Today's English Version*—Second Edition. Copyright © 1992 by American Bible Society. Used by permission.

Library of Congress Cataloging-in-Publication Data

Fuller, Millard, 1935–
   More than houses / by Millard Fuller.
   p. cm.
   ISBN 0-8499-3762-0
   1. Human settlements—Planning. 2. Housing—International cooperation.
3. Community development, Urban—International cooperation. 4. Habitat for Humanity International, Inc. I. Title.
   HT65.F85   1999
   363.5—dc21   99-24425
                                              CIP

Printed in the United States of America
99 00 01 02 03 04 05 06 QPV 9 8 7 6 5 4 3 2 1

Habitat for Humanity is building much more than houses. By building hope it is building relationships, strengthening communities, and nurturing families.

—Paul Newman, actor, Habitat home builder

CONTENTS

Acknowledgments

More than any book I have previously written, *More Than Houses* resulted from a team effort.

First of all, literally hundreds of Habitat affiliate leaders and other Habitat partners submitted stories and other material. I am deeply indebted to everyone who took the time to send in something. Much of it is included. Of course, I couldn't use every story, but I appreciate every story, idea, and suggestion. Every effort greatly enhanced the fullness and excitement of the book.

The writing of *More Than Houses* was done in several places over the course of more than a year and a half. I started writing at Debordieu, on the South Carolina coast near Georgetown, in a lovely beach house owned by our wonderful friend and Habitat International board member, Bonnie McElveen-Hunter. Writing continued sporadically over the following months in various places including a condominium on the Gulf of Mexico at Gulf Breeze, Florida, furnished to us by Gulf Breeze United Methodist Church, and at Glencove, the beautiful mountain cabin by the shore of Lake Rabun in north Georgia owned by our dear friends Chrys and John Street. My wife, Linda, and I also returned a second time to Debordieu for another week of writing. Finally, I wrote at our dining room table in Americus for several weeks toward the end of 1998.

I am profoundly grateful to Bonnie, Chrys and John Street, and the people of the Gulf Breeze United Methodist Church for their kindness and generosity in making those beautiful places available to me. And I am grateful to Linda for allowing me to take over our dining room for weeks, with paper piled all over the place!

I also want to express my gratitude to Linda for her help in translating my scribbled handwriting into a typed manuscript and for her many good suggestions concerning the overall manuscript. She has been a vital part of every book I've written, and I am enormously grateful to her.

My executive assistant, Sharon Tarver, also did much of the typing and retyping of the manuscript. She was ably assisted by Jane Phagan, Marsha Johnson, and Marilyn Ramsey.

Joy Roethlisberger, my special assistant, did a tremendous amount of work on the book. She served as the primary contact person with the book's principal editor, Lynda Stephenson, and Laura Kendall, the managing editor for the book at Word Publishing. Joy worked closely with me, Lynda, and the Word people on many aspects of the book and made sure everything moved smoothly on the project. I am enormously grateful to Joy for her effective work.

I want to thank Lynda Stephenson for another great job of doing the principal editorial work on the manuscript. This is the third book on which we have collaborated. I have so much respect and appreciation for Lynda, a true professional and dedicated Habitat partner!

Finally, I want to thank Joey Paul and his dedicated staff at Word. They are not only competent, but also a lot of fun to work with. Please know that I have a profound appreciation for all of you.

Introduction

Since the very beginning of Habitat for Humanity in 1976, I have written and published books about this house and community building venture. All of the books have been about various aspects of the incredible growth of what we do—building houses around the world.

My first book, *Bokotola,* was published in 1977. It told the story of the creation and first efforts of the Habitat idea at the small Christian community of Koinonia Farm near Americus in southwest Georgia and about its expansion in Zaire (now the Democratic Republic of the Congo) in Central Africa. That first book also articulated the core concepts that form the basis of Habitat for Humanity:

- Houses built for needy families with their full participation through "sweat equity"
- Sold to them at no profit and no interest
- Nondiscriminatory family selection criteria
- Modest but adequate houses constructed
- Neighborhoods built in conformity with our founding slogan, "A decent house in a decent community for God's people in need"

My second book, *Love in the Mortar Joints,* had as its principal theme the idea that we were putting God's love into action as we built houses and changed people and society for the better in the process.

The third book, *No More Shacks,* was written to boldly state the ultimate goal of our endeavor, which was to rid the nation and the world of poverty housing and provide at least a simple, decent place to live for everyone.

The fourth book, *The Excitement Is Building,* written in collaboration with my wife, Linda, described the phenomenal growth of Habitat across the United States and around the world.

The fifth book, *The Theology of the Hammer*, gave an in-depth look at the theological basis of this ministry, pointing out that true religion had to be more than singing and talking—true religion mandates action. It also stressed that, while theological, philosophical, and political differences tended to divide people, the "theology of the hammer" brought them together to drive nails, saw boards, put up walls and roofs, and do everything else that was necessary to build houses for and with needy families.

Finally, my most recent book, *A Simple, Decent Place to Live,* published in 1995, included a series of chapters presented as questions to give an interested person an overview of the entire growing movement of Habitat for Humanity: how the work began, why it has grown so rapidly, who the volunteers are, what the core philosophy is, why we work around the world, and so forth.

These books have been distributed in various ways by the tens of thousands. Some of them have been translated into several other languages. They have helped to spread the word about the work to more and more places and to increase understanding about this "hammering movement."

But I felt that another book, a very different book, was forming itself as we looked at Habitat's work, a book that illuminated how we are building much more than houses. The houses, to be sure, are incredibly important. A solid roof over the heads of families, sold on the "Bible finance plan" with no profit and no interest so that they are truly affordable to low-income people, is the visible and essential core of the ministry. As we approach and enter the new century, a new Habitat home is going up every thirty minutes. These houses are being built in more than fifteen hundred towns and cities in the United States and more than a thousand other locations throughout Canada, Mexico, the Caribbean, Central and South America, Europe, Africa, Asia, and the South Pacific. Habitat for Humanity is operating in more than sixty nations, and new countries are being added every year.

It is increasingly obvious, though, that the hundreds of thousands of volunteers, staff people, and homeowners across the nation and around the world are building *so much more* than houses.

We are building people.

We are building relationships.

We are breaking down barriers.

We are bringing people together.

We are promoting love and understanding.

We are building and revitalizing neighborhoods.

We are activating faith and planting hope in the hearts of people.

We are truly, I think, an exciting part in building the kingdom of God on earth.

In the following chapters, you are going to find stories that recount the transforming effect of Habitat on a great variety of people. You'll read about several intriguing surveys. One survey was done of selected U.S. Habitat affiliates and homeowners by a real estate and policy planning corporation under contract with the U.S. Department of Housing and Urban Development. Another survey documented the positive changes Habitat houses have brought in the country of Malawi in southern Africa. Yet another survey revealed improved health in an area of Zambia where hundreds of Habitat houses have been built. The findings of these surveys are interesting and very confirming of the positive impact of Habitat for Humanity on families and neighborhoods.

But, as stated above, this book is primarily stories of how the work of Habitat has had a life-changing impact on an incredible array of people.

There are wonderful stories of how the whole Habitat experience has transformed families around the world, making a tremendous difference in the lives of children, especially their academic performance.

There are stories of how health has been restored to people after they've moved into Habitat houses.

There are stories of how the building of Habitat houses and the influx of new families have revitalized neighborhoods and how involvement with Habitat has led to new careers or advancement in existing careers for many, many people.

There are thrilling stories of how Habitat has brought churches closer together, across all kinds of denominational lines, of how Habitat has helped build racial understanding and reconciliation, of how rich and poor, Democrat and Republican, Catholic and Protestant, liberal and conservative have been brought together through putting "hammer to nail."

There are inspiring stories of how people have been drawn closer to God through their Habitat experience. How the work is growing dramatically in prisons and having a powerful impact on people in those prisons who are building Habitat houses. There are also important stories of the involvement of college students and other young people who are the Habitat of the future.

And there are stories of romance! Literally scores of couples have found each other on a Habitat work site or in a committee meeting and have fallen in love and married. I routinely challenge these Habitat love birds to "go forth and make Habitots." A great many of them are doing just that.

More Than Houses is a book about impact, change, and transformation. As you read, I hope that you are informed and inspired, but I also hope for *more.* My fervent desire is that through these stories you will be caught up in this great venture of house building, home building, and people building and be brought closer to those around you and closer to God.

So be moved by these testimonies. Be moved to action. And if you are, never forget we need a vast array of committed people to "step up to the hammer" in this growing movement.

The families still mired in pitiful houses are counting on us to change their dreams into reality by being a part of building these houses—and so much more.

—MILLARD FULLER

CHAPTER 1

Making Dreams Come True

Every time I see a Habitat house go up, tears fall from my eyes because I know someone else's dream is coming true.
—SHIRLICE SPIVEY, HABITAT HOMEOWNER

A house is something visible. It is a place in which to be. An address—in a neighborhood. It is the site where loved ones live. It is where children study, play, and grow. It is where friends and family come to visit.

A Habitat homeowner in the western Ugandan city of Kasese once said, "There is nobody who is peaceful without a house."

A house is incredibly important to a family. A house is to a family what soil is to a plant. It is a place to be rooted, a foundation on which children can grow, develop, and become all that God intended.

But millions of people all over the world do not have a decent house in which to live. The United Nations estimates that as many as a billion and a half people live in inadequate housing. A hundred million people have no house at all. They are homeless.

Even in developed countries like the United States, Canada, the United Kingdom, or New Zealand there are thousands of families who have no decent place to live. In developing countries the numbers of families living in substandard housing are much greater.

We in Habitat for Humanity believe this sad situation is a disgrace. And we are committed to transforming that disgrace into grace. We believe every person, every family should have at least a simple, decent place in which to live. That's why our goal is to eliminate poverty housing from the earth.

We aren't the only ones who have noticed the problem and are working toward solutions. Eliminating poverty housing from the earth is also the goal of the United Nations Habitat program. In June 1996, national leaders from one hundred and seventy-one countries met in Istanbul, Turkey, for the second United Nations Conference on Human Settlements (Habitat II). I was privileged to be a plenary speaker at that conference. (See appendices for the speech and more on the U.N. Habitat program.) The Istanbul conference had two themes: "adequate shelter for all" and "sustainable human settlements development in an urbanizing world." The paper that resulted from the 1996 conference is called the Habitat Agenda. Government representatives, private sector people, nongovernmental leaders, and individuals from community-based organizations worked together to present a vision for affordable housing for all.

Dreams Unfulfilled

We've learned so much from our decades of work, not just about building houses, but about helping build lives for the people who will live in them. From abundant evidence, we know that families everywhere aspire to have a good place to live. When they live in miserable conditions, their thoughts turn to how to move into something better. They plan. They discuss. They talk about a way to change things. They dream.

But too often the plans, the discussions, the talk, and the dreams remain unfilled. And the usual reason for this unfulfillment is a lack of financial resources, a lack of guidance, and a lack of a helping hand from some person or organization.

Habitat for Humanity offers resources, a helping hand—or more correctly helping *hands,* and abundant guidance. And Habitat comes with the "Bible Finance Plan" that enables low-income families to afford the monthly payments for a decent and adequate house because the financing comes with no interest charged and no profit added. This plan is based on the clear Bible teaching not to charge interest to the poor. That ancient principle is God's formula for enabling low-income people to catch up.

More and more families all across the United States, Canada, Mexico, and around the world are seeing their dreams fulfilled by a Habitat house. The challenge is to greatly accelerate and expand this "dream come true" phenomenon to the point that there will be no unfilled dreams. We know it can be done. Some countries have already eliminated poverty housing, but the overwhelming majority of the two hundred nations of the earth still have great numbers of people languishing in poor housing. The Bible promises it can be done. With God, all things are possible. I believe that. We push ahead with the firm belief that God's Word is true and that the task God has given us can be fulfilled. It won't be easy, but what great challenge has ever been easy?

Your heart will be stirred and deeply moved by the following stories from homeowners whose dreams have come true and whose lives have been changed in profound ways by their Habitat house and by their loving and caring Habitat friends and partners. Who can resist a dream?

A Little Piece of Heaven

Judy Critchfield of Monongahela, Pennsylvania, was mired in poverty. Listen to her own words:

> I lived in one of those broken down little shacks that inspired the founding of Habitat for Humanity. Our house had no hot running water, no indoor bathroom facilities, and was heated only by a small coal stove in the center of the "living room" with blankets on walls, doors, and windows to keep out the cold. My husband and I were married in 1982. Between 1982 and 1994 we lived in eight houses and apartments. We were never allowed to have pets, and rarely were we permitted to do what we wanted to decorate or improve our home. We paid out thousands of dollars for the use of other people's properties and had nothing to show for it. With our financial struggles, owning a home was one part of the American Dream for which we had given up hope.

> Now, in answer to our prayers, God has given us the privilege of owning our own home, thanks to Habitat for Humanity. The immeasurable generosity of the churches and volunteers who worked with us was inspiring. One church, whose members did much of the work, also helped by giving us curtains and other furnishings and even Christmas gifts for our three children.

> We learned about the program in 1990 from a friend who deals in real estate. We put in our application and immediately began attending meetings and volunteering our time. In 1991 we were approved as prospective homeowners. Near the end of that year we started building our house. It took more than two years to complete the work, but in July 1994 we were able to move in. Now we have a

beautiful dog, our children all have their own rooms, and we have one and a half bathrooms. The house looks like a mansion to me. This home is not only a dream come true but a little piece of heaven.

No More Mildew

In north Georgia a divorced mother tells of her struggle and eventual victory over pitiful living conditions. She lives in Clarkesville, and her house was built by Habitat for Humanity of northeast Georgia.

"I divorced in 1983 after fourteen years of marriage. I thought I would never have a decent home for my son and me. The little two-bedroom house we were renting had no insulation. Mildew grew on most of the inside walls and some of the ceilings. I would scrub the mildew off, and it would come back. The one gas heater did little to keep us warm in winter. You could see daylight around one of the windows in my son's room. My son was ashamed to have friends visit," she said.

"I had been praying for some time for a better home, but it seemed doors slammed in my face everywhere I turned. One of my friends read about Habitat in the newspaper, and I decided to check it out. I found that if I was willing to work on homes for others as well as for ourselves, I would get a decent home. I learned how to put shingles on roofs, hang dry wall, and put in insulation. I painted, dug drainage ditches, and planted shrubs. I take great pride in my new home and in knowing that I helped others reach their dream, too. I am grateful to Habitat for the helping hand."

Shack Falling Down

On the other side of the country, in Chewelah, Washington, Dorothy Hinkey and her eight-year-old son, Steven, were living in a little shack that was ready to fall down. Dorothy said that their water pipes froze every winter, and they had to melt snow to wash dishes and take sponge baths.

She learned about Habitat for Humanity from an article in their local paper. An application was filed and the hoping, dreaming, and waiting began. With buoyant optimism, Steve saved his allowance and harvested aluminum cans. Dorothy saved every dime she had left over at the end of the month.

Then one day the call came informing Dorothy that she and Steven had been chosen for a Habitat house. Dorothy said she was not sure how she answered, but that she was sure she *yelled* her acceptance.

Dorothy and Steven worked diligently on the house until it was finished. Some nights, Dorothy said, she was so tired when she got home that she didn't want to move. But it was all worth the effort. She and Steven joyfully left the shack behind and moved into their dream come true.

Too Good to Be True

The Johnson family in Hinesville, Georgia, felt "stuck" with renting a house. Then Ms. Johnson saw a notice about Habitat for Humanity on the bulletin board of their church. They applied for a house. Ms. Johnson said she started dreaming that they were going to get a house. And her dreams came true! The Johnsons say they are blessed to have gotten help from Habitat for Humanity. Now they want to pass the blessing on to others.

J. D. Eslich of Cowan, Tennessee, says that their Habitat house has made a great difference in the life of his whole family. He said that when they were told they had been selected to have a house they couldn't believe it. "It seemed too good to be true," he exclaimed. "But it wasn't a dream; it was a dream come true." He said they worked every day on the house, and they were "so proud."

"Finally," he said, "the house was finished, and we didn't know how to act. We thank God every day for the difference this has made in our lives. It took time, work, patience, and lots of love. When it was all completed, it was our dream come true."

Taking Pride

T. L. and Angie Stepp of Erie, Pennsylvania, worked hard to get their Habitat house. They dug ditches, laid underground piping, poured cement, painted, and did many other jobs on the site. They gave up Christmas presents that year so they could pay off an accumulated debt.

T. L. and Angie painted the entire interior of their new home and completely landscaped the property. They continue to work on other Habitat houses.

T. L. is effusive in talking about their new Habitat home: "When you are a homeowner, you feel like planting grass and flowers to make it look nice. There is something about the sound of 'mine.' When I am coming home from a stressful day at work, I know the warmth and comfort that awaits me at our home. I sit down and admire my home—it was a dream come true. We love our house! I thank God for Habitat and the people who work for

Habitat. I have made lifelong friends, and our family has been made stronger. I will never forget what Habitat has done for us."

The leaders of the local affiliate of the Stepp family, Greater Erie Habitat for Humanity, say that they are a model family (a mom and dad and two fine boys)—a true success story. Both T. L. and Angie speak for Habitat in churches and continue to support other families working through the program.

The six members of the Del Valle family of Palms, Michigan, was living in a small two-bedroom mobile home. They moved from that cramped place to a three-bedroom Habitat house with a big backyard in a quiet neighborhood. Mario and Elaine said that they had always dreamed of having their own home, and now that they are proud homeowners, they "guess that dreams really do come true."

Dennis Lehman, president of Habitat for Humanity of Johnson County, Texas, submitted an unusual and heartwarming story in regard to the building and dedication of their first house in 1996. "One Saturday while the men were cutting tree branches and doing the heavy work of clearing up the lot where the house would be built, my wife was picking up small trash. She came across a little rubber doll. It was dirty, naked, and looked pretty bad, but the doll did have all its body parts, including some very matted, disheveled blonde hair. She took the doll home and started cleaning it up. After several bleachings and scrubbings, it was ready for a new outfit of clothes and a trip to the beauty parlor. The little doll actually looked real good all clean and shiny and dressed fit to go to church," he said.

"At the dedication service after the passing of the Bible and just before the keys were given to the homeowners, the doll was presented to them with a brief explanation of where she came from. There was hardly a dry eye in the crowd when the following poem written by my daughter, Jennifer, for the occasion, was read:

ONCE ABANDONED

I was left naked and all alone
On just a foundation without a home,
I was cold and feeling desolate
Never before had I been so desperate.

Then one day things began to change
At least I saw a hope for something to gain,
They came out to clean up the mess
And I was bathed and given a dress.

The garbage that was left on the lot
Was disappearing and materials were bought,
To build a new home for a family to live
After much hard work, it is now time to give.

Now I'm seeing my dreams come true
No more feeling of sadness and blue,
Because I've been given a family and home
I will be loved and no longer alone.

Two Prayers

Lavenna Wackler's "dream come true" story started with two prayers. She and her husband, Steve, and their two children lived in Piqua, Ohio, in a small apartment. First, Lavenna prayed for a bigger place: "All right, God, I don't know how it will be accomplished, but please, just a little larger place, maybe with a yard for the kids to play and a few plants—tomatoes and flowers would be nice." Next she prayed for Habitat for Humanity to come to their local area: "Please God bring this organization here. There are people who need help."

Soon thereafter, Lavenna was working at her church and talking with the pastor, Charla Koener. She told Lavenna there was to be a meeting at the church to organize an affiliate of Habitat for Humanity. Lavenna went to that meeting and became heavily involved in the work. Eventually, she applied for a house for her family.

The acceptance letter arrived on Christmas Eve 1992. The groundbreaking was scheduled for May 8, 1993. By Christmas of that year, the Wacker family was ensconced in their new home with a yard, trees, and an affordable mortgage. Lavenna said that if there had been an unlimited budget to plan and build with, their house could not have been better. The family was full of giving glory to God for the great good fortune in their lives.

A Mansion

Shirlice Spivey is a Habitat homeowner in Myrtle Beach, South Carolina. "My children had this dream while I prayed for the dream to come true," she said of the process of being accepted for her Habitat home. Now she says, "Every time I see a Habitat house go up, tears fall from my eyes because someone else's dream is coming true. This is why I fight so hard for Habitat, because they are doing the works and wonders of the Lord. I may

Shirlice Spivey and her friend Al clear the lot for her Habitat home in Conway, South Carolina.

not live in a mansion, but my house is like a mansion. I have a mansion from the Lord because of Habitat and the First Presbyterian Church."

Shirlice's ten-year-old son, William, wrote the following about their Habitat house: "I am proud to say that Habitat made my dream come true. I always wanted my own house. I told my mom that when I grow up and become famous like Michael Jordan, I'm going to build us a home with a big kitchen for her, because she loves to cook, and a big yard to play football and basketball in. I'm not famous, but I have my dream house thanks to Mrs. Judy Swanson and Habitat for Humanity and, most of all, God, because He heard me from Heaven and sent Habitat to my family and me. My friends ask me how they can have a home and a dream to come true like mine. If someone would help, they could have a house. There are lots of children asking for help to get a Habitat house, so someone please listen. The most important thing about my house is that I helped build it, me and my family."

A Literal Dream

Gloria Putiak of Habitat for Humanity of Broward County, Florida, head-quartered in Pompano Beach, shared a wonderful story about a woman who literally dreamed about someday owning a home. "One night, before she had even heard Habitat existed, Vickey Burke dreamt that she was applying for one of several homes being built by a group of people. A woman handed her an application, prayed with her, and brought her to see the homes in the process of being built from cement slabs. Vickey was told her home would have four bedrooms."

Within a year, Vickey saw a television commercial explaining Habitat for Humanity, and she called for an application. Soon she was one of six families selected for the Harmony Village Blitz of 1996, and her home had four bedrooms to accommodate her six children. "I was privileged to be present at the Harmony Village Blitz and spoke at a big Habitation service at the end of the week," wrote Gloria. "Vickey also spoke on that occasion. She was and is an inspiration. She gives thanks to God for the fulfillment of her dream. When she prayed to God to show her if the dream was truly of Him, she felt directed toward a verse in the Bible: 'Write down the revelation and make it plain on tablets, so that a herald may run with it. For the revelation awaits an appointed time' [Hab. 2:2–3]," said Gloria. "Vickey took this to mean that she should tell people about her dream so when it became a reality they would know it was really of God. She has been vocal in her neighborhood about her dream coming true through Habitat."

Peace of Mind

John Cerniglia, executive director of Cincinnati Habitat for Humanity, shares the incredible faith and perseverance journey of Pedro Alicea and his family. Pedro was born in Puerto Rico and had dreamed of a home of his own for most of his life. But the dream was long deferred because Pedro and his family were poor. Furthermore, he was burdened for years with medical bills connected with his wife's kidney dysfunction. But Pedro always had faith. The family's credo, repeated three times weekly at services they attended at Norwood Pentecostal Church, is "The Lord will provide."

The family was living in a three-bedroom apartment with a monthly rent of $420—almost half of his salary. The neighborhood had also deteriorated with frequent gunfire and drug deals.

In August 1994, the Alicea family—Pedro; wife, Janette; and children, Michelle, Eunice, and Pedro Jr.—were chosen to receive a Habitat house. When Pedro got the good news, he exclaimed, "This is a blessing the Lord is providing us. It's a dream that is coming true. You've got to believe in miracles."

Work started on the house in September and continued for eight months. Then vandalism started. Some neighborhood kids broke two exterior light fixtures, kicked in the front door, and poked holes in the dry wall. When asked about the damage, Pedro calmly replied, "I know the Lord will provide peace. That's why you buy a house. To have some peace of mind." The vandalism stopped, and the damage was repaired.

The joyous dedication service for the Alicea house was held on July 22, 1995. Their pastor, Donald Helton, offered the invocation, in which he observed, "A lot of hands can build a house. But only love and communication in the family can build a home." Pedro said simply, "My dream has come true."

John Cerniglia reports that the Alicea family is doing great as homeowners. He says their mortgage payments are always on time (about half of what they were paying in rent), and they have improved the property with a fence, a small above-ground pool for the children, and a deck. At Christmastime they put up lights, and in the summer they plant beautiful flowers.

Everyone Needs Help Sometime

Another happy Habitat homeowner family is the Justice family of Lake Junaluska, North Carolina. Terry Justice explains her feelings about the house: "When we first moved in, it was like spending the night with someone else. I couldn't really believe the house was ours. It is a strange but wonderful feeling."

"Every day when I walk through the house doing my work, I'm reminded of everyone who helped on our house. It's an amazing feeling to know how many good and decent loving people there are. It's like one huge family helping someone who needs help.

"Having a Habitat house is great, not anything of which to be ashamed. Everyone needs help sometime in their life, and I'm sure glad Habitat was there for us. We now have a warm and decent place to live. We are so proud of our Habitat house and all our Habitat family."

I Will Remember You

In 1997, Habitat for Humanity of Williamson County in Franklin, Tennessee, built a house with and for Connie Martin. Connie is a single mother with two teenage sons. Her mother also lives with her. The house was sponsored by Maxwell House Coffee and a Sunday school class at Brentwood Baptist Church. The volunteers who built the house were members of the Sunday school class. Connie became friends with many members of the church and was introduced to the congregation at a Sunday morning service just before she and her family moved into their new home before Christmas 1997. The following are excerpts from a letter Connie wrote to the class.

"It is hard to find the words to express what I feel today toward all of you. It is not often that I am able to become acquainted with such a wonderful group of Christian people. Mere thanks, somehow, does not seem adequate, and I don't know if there is anything I could do to recompense you for the generosity of your hearts," she wrote. "I do know that I will remember you each time I sit on my wonderful sofa or chaise lounge. As I sit at my dining-room table, I think of people who really understand what Christianity is all about. Every piece of vinyl, insulation, woodwork, every stroke of the paint-brush serves as a reminder of the kind, caring people who chose to be a part of our lives.

"To emphasize how wonderful it is to have a warm, comfortable place to call our own, I wish to share a story that happened several years ago when I was about ten years old," she wrote. "As a family of six, we lived in a tiny one-bedroom house that was heated only by wood. There were so many holes and cracks in the ceilings and walls that daylight could be seen. I remember my precious mother arising early each morning to start the fire for the rest of us. I especially remember when it was so cold that I was unable to sleep, she would take a towel and warm it on the wood stove and then wrap my feet in it to keep me warm. This act of love and kindness I have never forgotten, and I promised myself that when I was older I would provide my mother with a warm, comfortable place to live. I finally am able to fulfill that promise thanks to you. I feel that I have been blessed not only with a home, but with wonderful friends who have lifted my spirits. I can only pray that God will richly bless each and every one of you."

Christmas Present

Tony and Yvette Brown of Chester County, Pennsylvania, are another Habitat family that moved into their new Habitat house just before Christmas. Their exciting story ends with a new twist on "Christmas present" as told by a volunteer, Jason Pyrah.

"It was a few days prior to Christmas 1996 when I visited Tony and Yvette. As a volunteer I had worked numerous hours on their house and became friends with them. The house was completed in time for the holiday season and they were so thankful. The beautifully decorated Christmas tree was in place in the corner of the living room. Yvette, who never stops smiling, asked me if I noticed anything different about the tree. I replied that it looked like a normal tree, complete with lights and ornaments. She told me there were no presents under the tree. I just thought maybe they were not out yet. Yvette told me that the present was *all around* the tree. The house was the Christmas present! It was at that moment that I truly knew the value of a volunteer's hours. Now that I am the volunteer coordinator for Habitat for Humanity of Chester County, I never tire of telling my story to other volunteers."

Must Be a Trick

Cherie White of Ephrata, Pennsylvania, long dreamed of a house of her own. But when she began to make inquiries, she was rejected again and again. She often thought there was no hope for her family. Then she discovered that Lancaster Area Habitat for Humanity was going to build a house in the Ephrata area. Cherie thought to herself, *Don't stop and inquire; don't even bother. You can forget it!* But she did inquire. She went through the application process, and when she was told she had qualified she thought maybe it was a trick. She started putting in her sweat-equity hours. All the time she kept wondering what would go wrong. It took her a long time to believe that sometimes dreams really do come true.

Cherie completed her five hundred hours of sweat equity as her dream house was built. Today she and her family are living in that dream house in Ephrata. Here are her own words: "Habitat for Humanity gave me an opportunity to fulfill my dreams. They make housing affordable to low-income families. . . . They make dreams come true. Habitat didn't give me a house. . . . Habitat gave me an opportunity to work hard to achieve what I wanted. I gained valuable knowledge and experience as I met the requirements for homeownership. Going through the process helped me to know myself better."

Getting Out of the Projects

Mary Mickens of Louisville, Kentucky, spent twenty-four years in public housing. She long dreamed of getting out of "the projects" and into a home of her own. In 1997, her dream was realized. She worked on her house for eleven months. She helped dig the foundation, put siding and dry wall on her house and other homes, painted, and helped in many other ways. In November, she moved in. She paid $281 a month for her apartment. Her

Habitat house, which has more space, costs her only $212 a month. When asked about her new home, Mary replied simply, "I love it!"

No One Peaceful without a House

Habitat for Humanity is doing extensive work throughout Central America. Thousands of homes have been built in El Salvador, Honduras, Guatemala, Costa Rica, and Nicaragua. Here is a typical story of a Habitat family in that part of the world:

"My name is Ann Luz Arauz Pineda. I am from the city of Jinotega, Nicaragua. I work for the city. I have four small children. Right now I am living in San Isidro—a neighborhood in Jinotega. It was hard for me and my children when we didn't have our house. I felt very sad and often wondered when I was going to have a comfortable house where I could feel at peace, relaxed, and without problems. My prayers to the blood of Christ were not in vain, but I suffered many difficulties and serious problems with patience," she wrote.

"One day I was at work, and a friend urged me to sign up for a Habitat project that would soon begin building in my neighborhood. For me it was a light illuminating my life. I began with all my spirit and faith in the Lord. I did everything possible to make it to the meetings and to work with all the other homeowners. At the beginning it was very hard, but within the group I learned, 'Everything is possible if we are all united in the love of God.' For me, it was impossible to build my house on my own. I can testify how God helped us individually to get ahead, conquering all the problems and difficulties that we encountered during the eight months constructing the sixteen houses in our project. Thanks be to God!"

The Reverend Ivan M. Rusenge, diocesan secretary of the South Rwenzori Diocese of the Church of Uganda, submitted the following stories from Habitat homeowners in Uganda, telling of how their dreams were realized in getting a good place to live through Habitat for Humanity.

Abdu Baluku Musema-Sivi said that he and his family were living in a single room in Kasese: "So I was very happy when I got a Habitat house at the Kamaiba Habitat site in Kasese. We have planted trees, pawpaws, mangoes, flowers, and banana plantations. I am quite fine with my family now. We have completed the mortgage, and we have built four more rooms on our plot. We thank you for the great assistance which was rendered to us. May Almighty God encourage you to assist many poor people without shelters or accommodation in the world as there is nobody who is peaceful without a house."

Carrying Stones on Their Heads

Mbusa Bwanandeke in western Uganda conveyed a great story of his experience with Habitat. "On December 20, 1995, my family happily occupied our new house on Bulunga Hill in the Rwenzori Mountains. The house had been constructed by the family, especially my wife, Mbambu, and her three little children who carried stones on their heads up and down the hills. After moving into the new house, our baby boy was born, so our family is now made up of four children who are happy at the sight of their new house. They love the house and demonstrate this love by planting flowers around it and cleaning inside and all around the house. Their mother is joyful because she is assured of shelter and comfortable rooms for her children."

"As a civil servant who gets monthly pay below the living wage of a human being and saving very little, I would not have been able to construct a permanent house on my own, even if I worked up to retirement. However, with a loan from Habitat, combined with local abundant building resources, meager savings, and family labor, we now have a home. And we have forgotten the mud and waddle grass–thatched housing conditions attached to my father's homestead."

Finally, Wilson Baswikira writes the following touching account of his dream come true: "I had no shelter. I was staying in one room with my family of eight people. All of our belongings had to be in that very room. Then I was visited by the Reverend Ray Cunningham, who was an employee of Habitat for Humanity. He saw how we were squeezed into that small room, so he promised that I would be given a chance to build a house with an interest-free loan and allowed to repay the money over twenty years. I agreed but with a doubting mind. I only said, 'Let Your will be done.' Within a few months I saw a reality in my life of having a good block house in which I am now living happily with my family."

"Economically, we are working hard so that we can finish off the mortgage repayment and own this house permanently. Socially, we feel comfortable in our own house and feel free with our visitors because we have enough room to host them. Politically, my family is recognized as a family with potential because we are a happy family. We have seen an exciting reality in our lives. My family is very healthy because we are staying in a decent house."

Neighbors Again!

The dreams continue, sometimes with the most wonderful twists. For Monica Gutierrez in Watsonville, California, realizing her dream of getting a Habitat house also meant being reunited with a former neighbor, Anna Herrera.

Monica's friendship with Anna began in 1967 when they were neighbors on a Brussels sprout farm in California. Eventually, they went their separate ways and never dreamed they would see each other again. Monica went back to Mexico for a while, was married there, and then returned to California with her husband.

Once again, she was living in the same area as Anna, but they weren't neighbors. Both of them applied for a Habitat house with Habitat for Humanity of Santa Cruz County. They were both selected to have Habitat houses at the same time—Mother's Day 1996. And they moved into their new Habitat houses—side by side—at the same time in February 1998.

I Can Dream Big!

Diane Johnson, a happy Habitat homeowner in Annapolis, Maryland, writes eloquently about her Habitat house:

My husband and I were approved for a Habitat house, and not only did I have dreams but they began to come true! Watching our house going up and meeting some of the most wonderful people was a joyful experience in itself, but the end product was beyond anything I could have imagined for my family. Our home! Our very own home! The feeling is beyond description for me. Waking up in the cold mornings with heat that's actually staying inside. Being able to take a shower and have enough hot water to finish it. Baking a birthday cake in an oven with temperature controls that work, and having a floor level enough to have a perfect cake. We have enough electric outlets so I don't have to use extension cords and worry about a fire. There are so many little things that many people take for granted that are so wonderful to my family now.

My husband loves the yard and keeps it looking great. One of the first things we did was to plant two dogwood trees in memory of my mother-in-law and father. We will be here to watch them grow, along with the flowers I receive from children and grandchildren. I'll never

Michelle Roberts of Americus, Georgia, looks forward to relaxing in her new home. Photo by Robert Baker.

have to leave a plant behind again. That means more to me than just about anything. I now have roots!

Has my life changed outside the house? You better believe it! My husband returned to teaching and went back to school. One son has graduated from high school and is making plans for college. The youngest finally has his own room and loves his privacy. He now has a quiet place for home-work and is on the honor roll at school.

God has been good to our family, and I thank Him for touching so many people who volunteer for Habitat and making them compassionate enough to help others. Now that I know how to dream and that dreams can come true, I have another. That dream is that three times as many families in the next ten years, as in the first ten years, will get new homes thanks to Habitat and its wonderful volunteers. I can dream big!

Henry and Mattie

Lee Hoover, former president of Southeast Volusia Habitat for Humanity, New Smyrna Beach, Florida, submitted another heartwarming story of a reunion of a closer and more personal kind: "Henry Smith is the most popu-lar Habitat homeowner in our community. He was over sixty-five and had only one arm. He worked at a local gas station, so he knew everyone and everyone knew him. Their home had fallen into such disrepair that the city had condemned it. His wife, Mattie, had moved to Titusville to stay with some family members there, but Henry continued to live there illegally. He had nowhere else to go. I learned that Henry's house was going up for auction, so I asked our treasurer to go to DeLand and bid on the property. He did so and was the successful bidder. During this time, I had to appear before the zoning board to get permission on several issues. Mattie was there. That's when I saw her for the first time. She threw her arms around me and exclaimed, 'You are going to save us, aren't you?' That was a first, and nothing else has ever happened since to give me such a feeling."

The affiliate built a house with Henry and Mattie so they were able to be reunited. Because he was such a popular man in his community, the local media gave extensive coverage when the Smith house was dedicated. Henry worked diligently on the house. In fact, he went well over the required sweat-equity hours. And he continues to work on other Habitat houses.

Darkest Nightmare

Of course, not everything goes right all the time. In fact, the worst can seem to happen, turning dreams into nightmares. Even then, being a Habitat owner can offer something extra, something more than houses. The story of Ora Moore is such a story.

Ora Moore, a Habitat owner also living in New Smyrna Beach, was charged with murder. Ora was a single mother of two sons, and her mother moved into their three-bedroom Habitat house. Soon thereafter, her youngest son was murdered. With little evidence, the state's attorney charged Ora with the crime. The community in which she lived deserted her entirely. They felt she was guilty, having done the dastardly deed in a rage brought on by alcohol or drugs.

Lee Hoover and other Habitat officials, however, stood by Ora. I have never been more proud of a Habitat leader than I was and am of Lee Hoover. She said that she and other Habitat people had frequently stopped by to visit Ora after the family had moved into the house, and none of them had ever seen any signs of alcohol or drugs. Ora was eventually brought to trial for the murder and found innocent. Lee then wrote a lengthy letter to the editor of the local paper further supporting this beleaguered woman. Lee and her fellow Habitat volunteers made the difference for this courageous woman in her darkest nightmare hour.

The man in Uganda was perceptive: "Nobody is peaceful without a house." Dreams are coming true in more and more places through the ministry of Habitat for Humanity. And even when the dreams seem to sour, as in the case of Ora Moore, there are brave and loving people like Lee Hoover to set things right and restore the dreams. I thank God for the dream builders and for the dream restorers who offer so much more than houses.

Incredible, unbelievable changes occur when people have homes of their own. And many incredible, unbelievable stories follow to prove it.

CHAPTER 2

Enhancing the Lives
of Children

*You don't have to have money to be happy and raise healthy kids, but
you do need a safe, clean, dry home—a home kids can bring their
friends to and not feel upset because of its condition—
somewhere they feel safe.*
—Marilyn Gilbert, Habitat homeowner

Perhaps the most dramatic change I've seen over the years is in the children of Habitat homeowners. Healthy self-esteem, honor rolls, high school diplomas, college hopes. These are the words that fill the stories that come to me year after exciting year. The impact of a simple, decent place to live on child development has always been very obvious to me. Now, finally, a government study proves it.

In 1997, Habitat for Humanity International and certain selected Habitat affiliates in the United States collaborated with Applied Real Estate Analysis, Inc. (AREA), a real estate and policy planning corporation headquartered in Chicago, to conduct a study to learn about the homeownership experiences of Habitat homeowners. The study was commissioned by the Office of Policy Development and Research of the U.S. Department of Housing and Urban Development (HUD).

The objective of the study was to learn about the experiences of Habitat homeowners directly from the homebuyers themselves. The information was obtained primarily through two research methods: interviews with Habitat homeowners and focus-group sessions. A total of ninety-five in-person interviews and thirteen focus groups were done, involving nineteen affiliates that had built a thousand houses. The nineteen affiliates were scattered all across the country, from Washington, D.C., to Fresno, California, and from Milwaukee, Wisconsin, to Jacksonville, Florida.

In regard to children, this extensive study concluded:

Homeownership seems to be having an overwhelmingly positive effect on homeowners' children. Parents emphasize the feeling that their family has stabilized. It was typical for homeowners to say that their children now had privacy and a door to close. One owner's son wrote a school essay on what his new home meant to him. Many respondents said that their children were home more often and brought friends home to visit and spend the night—something they had not done before because their house had been too small or the children had been ashamed of its condition. One homeowner mentioned that rats had infested her former home; she said her children were too embarrassed to let anyone visit and refused to tell people where they lived. Another homeowner cried as she told of being able to give her daughter a sleep-over birthday party for the first time; before moving into her Habitat home she and her two kids had shared one bedroom and the living room couch in her mother's home. Homeowners' children piped up during the interviews to show what part of the house "they had built" and emphasized that it would be theirs one day.

While the study did not quantify the impact on school performance, it did report that "a number of households attributed their children's improvement in school to the increased stability they enjoyed as a result of owning a home."

My Boys Are No Longer Ashamed

Several years ago, we built a Habitat house for a very poor family in Americus, Georgia. The new house was built right behind the derelict shack the family had been occupying. The old house was literally falling in on the single mother and five children. When the new house was finished, the old house was demolished.

Soon after the house was completed, I took some visitors to see the family and their new house. In the course of the conversation, I asked the mother, "Annie, what is it about your new house that means the most to you?" She shot back immediately, "My boys are no longer ashamed for their friends to know where we live!"

Think about that. What damaging effect does it have on children to be perpetually ashamed and embarrassed about their home? I am convinced that the loss of self-esteem and pride brought about by living in poor housing negatively impacts children in all areas of their lives, particularly in their schoolwork. But moving children into good homes has a tremendously positive effect on them.

First Success Story

Consider the Johnson family who moved into the very first house built at Koinonia Farm, the birthplace of the Habitat ministry. The father of that family was a very intelligent man but had never had a chance to get an education. He was illiterate. I closed the sale on his house, since I was still practicing law at the time. He couldn't sign the mortgage so he just put his mark, an X.

The Johnson family had moved from an uninsulated tarpaper shack into a modest but solid house. They raised their five children in that house. As this book was going to press in 1999, the parents still live in the house. The children are grown now, and all are successful. One daughter has her own law firm in Washington, D.C. For several years, she served on the board of Koinonia, the organization that built a house for her family when she was a little girl.

A young boy in Mexico City proudly displays his new Habitat house. Photo by Robert Baker.

Valedictorian and Full Scholarship

Juan Garcia and his family were chosen to receive one of the first houses built by Immokalee Habitat for Humanity in southern Florida. Immokalee was the first Habitat affiliate in Florida and the third in the nation. The two children in that family blossomed in the following years. The oldest son, Noel, was valedictorian of his high school class in 1995 and went to college on a full scholarship.

Little Interviewer

In the northern part of that state, in Pensacola, the Stapp family moved into their new Habitat house in 1995. I was in Pensacola in September 1996 to dedicate the fifty-thousandth Habitat house built worldwide. While there, I was approached by Shawna, the ten-year-old daughter of the Stapp family. She wanted to interview me for a school project. She was very polite and quite professional. Following the interview, she stood up and extended her hand to me. "Thank you for the interview. I want you to know that Habitat for Humanity has changed my life. I was doing very poorly in school before we moved into our new house. Now I am an 'A' student."

Betty Salter, director of Pensacola Habitat for Humanity, confirmed what Shawna told me and further reported that the entire family had blossomed, including becoming active members of Gulf Breeze United Methodist Church, the sponsor of their house.

My wife, Linda, and I were back in Pensacola in March 1998. On Sunday morning, I preached at the Gulf Breeze Church. The Stapp family was in the congregation, and I had the joyous experience of recognizing them from the pulpit. That afternoon we helped dedicate a new Habitat house in the Stapps' neighborhood. Following the service, we walked up the street to their home. Shawna, now twelve, proudly showed us her beautifully decorated room. She continues to do exceedingly well in school and in other areas of her life.

Children Teach Their Father

Yet another remarkable success story emerged in Florida, in the city of Sarasota. I learned about it a couple of years ago when I was there for some speaking engagements. My good friend and Habitat International Board member John Schaub was escorting me as I made my rounds of various engagements.

While we were driving along, John looked at his watch and exclaimed, "Millard, we've got a few minutes before your next engagement. I want you to meet a Habitat family." He promptly pulled into a driveway. We got out and rang the doorbell. Within seconds, the family was at the door—a mom and a dad and three young daughters.

We visited for a few minutes in the living room. The family was so excited to see John. They were obviously good friends. The girls wanted to show me their rooms. So we walked down the hallway to the back of the house. The girls were very proud of their rooms and the certificates of academic achievement that were neatly displayed on the walls.

Later, as we drove away, John told me the family's remarkable story. The father was a Vietnam veteran. Upon returning home after his tour of duty, he did not do very well. His family was poor, and they lived in poverty housing. The girls were failing in school. Somehow the family learned about Habitat for Humanity and were chosen to have a house. At the time of our visit, the family had been in the house for four years.

Almost immediately, the girls started doing better in school. The oldest daughter did so well that she started talking about going to college and medical school. The father, unfortunately, was in poor health. His condition was exacerbated by his cigarette habit. One day, while driving out in the city, he suffered a heart attack and almost died. That experience caused him to quit smoking. He said, "I want to live to see my daughter graduate from medical school."

The father was also illiterate. He started studying with his daughters and learned to read and write! That family continues to grow and blossom.

What Soil Is to a Plant

As I stated in chapter 1, a house is to a family what soil is to a plant: It's a place where a family can put down roots. The children especially feel safe and secure. In that environment, they are able to grow and blossom into all that God intended them to be.

That is certainly true of a little boy named Charlie. Jerry and Cindy Schultz told me this exciting tale when Linda and I were in Washington in 1995. We were visiting them in regard to their long walking trek across the United States in 1996 to attend the twentieth anniversary celebration of Habitat for Humanity that was held in Atlanta. Jerry and Cindy were on the family selection committee of their local Habitat affiliate. They received an application from a family that had been living in a car. The family had moved to a rundown apartment, but their living situation was still deplorable.

After due consideration by the committee, the family was chosen to have a Habitat house. Jerry and Cindy went to their apartment to deliver the good news. After visiting for a few minutes, the magic words were spoken about their new Habitat house. Little ten-year-old Charlie had been listening intently to the conversation. When he heard that they were going to get a house, he started jumping up and down, yelling, "We won! We won! We won!"

In the coming months, the house was built. Charlie helped a lot, and he even built a tree house in the backyard. Finally the family moved from their rundown apartment into the new home. Charlie was most excited of all.

The new house was in a different school district. Somehow Charlie's school records were not transferred to the new school, and he was placed in a regular class. In his previous school, he had been classified as a slow learner and placed in a class of slow learners. The new school didn't know this. The "error" was not discovered for six months. By that time, Charlie was a strong B student. What made the difference? The school officials could only conclude that Charlie had moved into a new home and had received a lot of love and affirmation. Those changes in his life made an incredible difference in his performance at school.

From Australia to Canada

The story seems to be the same all over the world. Sonja is a young mother in southern Australia. When her husband deserted her and their four children, she was forced to rent inadequate housing. The monthly rental payment consumed more than half of her meager income. In five years, the family moved six times. The eldest son developed serious behavior problems and performed poorly in grade school. However, when the family moved into a Habitat home, the boy's behavior improved, and he went to the top of his class academically.

Wilmer Martin, president of Habitat for Humanity of Canada, shared about the impact of a house on the children of the first Habitat homeowners in Kelowna, British Columbia. "Jim and Debbie," he wrote, "worked very hard at low-paying jobs in order to care for their family of three children. The house they lived in was condemned. Friends encouraged Jim and Debbie to apply for a Habitat house. They did so and were accepted. After they moved into their new house, I was invited to have tea with the family. Jim and Debbie showed me through their home. They told me, 'We will have this house for a long time. We want our children to be proud about where they live and to bring their friends to their home.' The children showed me their schoolwork. Their daughter told me she was on the principal's honor roll and president of her church youth group. Their son, who was having trouble learning previously, was also on the honor roll. Their oldest daughter said she's going to be a surgeon some day. She plans to attend a nearby university to fulfill her goal."

Nobody Liked Them

Marilyn Gilbert of Provo, Utah, wrote movingly about her family's previous living situation and about how their new Habitat house has made such a difference in the lives of her children:

My story begins in 1992. I heard from a neighbor about a new affiliate of Habitat for Humanity starting. I became involved almost from the beginning. I did what was needed just so the affiliate would stay working. At that time, my husband and I, with our three children, ages eleven, thirteen, and fifteen, were living in a ten-by-fifty-foot mobile home. It was more than thirty years old and had more holes in it than a sieve. The roof leaked, the plumbing was bad, and the cupboards in the kitchen were falling apart. It had two small bedrooms, so we put the kids all in the same room. When our son became too old to sleep in the same room as the girls, he slept in the front room. It was three years before I found out we were getting a home. (It's not only a house, it's so much more; it's life, love, and a second chance.) Anyway, it took a year to complete, and the difference it has made is almost unbelievable. Our children are much happier, and are now going to school.

Before they said that no one liked them and they had no friends. They didn't want to bring anyone home with them because we lived in a trailer park, and everyone knows that only drug dealers, people on welfare, and trash lives there. My youngest is now sixteen and will graduate from high school before she is seventeen. She is now a full year and a half ahead of others her age and is a B-plus student. She plans to go to college and would like to be a doctor. This has made such a difference in her life. She has gone from a quiet, shy, self-conscious person to one full of life, someone who wants to see and do everything. I am so very proud of her, as I am of all my children.

You don't have to have money to be happy and raise healthy kids, but you do need a safe, clean, dry home—a home kids can bring their friends to and not feel upset because of its condition—somewhere they feel safe. Kids are always at our home now, because it is a place where there's love and my children are happy.

Learning to Be a Future Builder

Beth Otto of Cataldo, Idaho, has a story to tell about the positive impact her new Habitat house had on her family and especially on her oldest son, Drew. Beth wrote not only about the house but the whole process of building it with their local affiliate, Silver Valley Habitat for Humanity. On the anniversary of the family's moving into their new Habitat home, Beth wrote: "Drew was twelve when we began dreaming of a house with Habitat. He was old enough to appreciate the phases of decisions, design, and development. Because we homeschool, he was always available to help and learn 'hands-on' skills.

"For Drew this was an 'up close and personal' way for him to evaluate vocational choices within the security of our loving affiliate family. The many talented 'professionals' taught, challenged, and encouraged him to learn not only construction skills but people skills as well. As a result, Drew has grown not just in ability but also in self-confidence. He believes that he would like to do framing construction work in the future. This has given us direction for his course of academic study and opportunity for his continued involvement in Habitat," she added. "Drew is fourteen now and is very helpful with 'around the house' projects. His unlimited potential has been tapped and cultivated by the building of our new lives."

Pat Shondrick, executive director of Habitat for Humanity of Southern Portage County, Ohio, told about a young man whose career path was impacted by his work with the local affiliate. "Russell did not do well in school. He had to attend summer school because he had failed English. The year his family moved into their own house he applied to Maplewood Vocational School. Even though his grades were not good, he asked to be in the carpentry class. He had to write a paper explaining why he wanted to be in the class. Russell told how he had helped build five Habitat houses, one of which was his own home. On the basis of his paper he was accepted into the carpentry class.

"On the first day of school his mother got a call at work from Russell, telling her how much he liked school and how he couldn't wait to go tomorrow. This was the first time he had ever liked school. When his mother went to a parent-teacher conference she learned that Russell was getting an 'A' in English. His carpentry teacher told him he would be tougher on him because he knew he could do it. All of his teachers had glowing reports on his work. Habitat for Humanity does more than build houses," Pat observed. "We build new lives for people."

Missi Tells Us

The children themselves often make a point to write how much their new house has changed them. Missi Christensen, daughter of Habitat home-owner Susan Christensen of Grand Island, Nebraska, wrote eloquently about the power of her family's getting a new home and what it meant to her and the whole family.

I was eleven years old when my mom, Susan, my older brother, Brian, and I were chosen as a recipient family for the third Habitat house to be built by Grand Island Area Habitat for Humanity. It was my last year in elementary school, which meant that the coming year was full of big changes and expectations and a whole lot of growing up. This was also the year construction was scheduled to start on our house. We'd already helped on the second house, and we were anxiously awaiting "our turn." But bad weather and organization problems delayed us for another year.

Okay, great. Here we are, ready to build, and winter sneaked up on us. But the building was underway. Board meetings and family nurturing meetings kept us involved throughout the cold months, and Tom and Peggy Beswick were chosen as sponsors to help us with anything we needed. Habitat members gradually became acquaintances and eventually became my friends. Someone was always at my concerts and games, and everyone beamed with pride when I proudly displayed my straight-A report cards. They were also there to help me throughout my confirmation process at church. Though we hadn't even broken ground yet, I'd built a dwelling of friends that would shelter me for life.

Finally, the ground thawed and building began. Teachers, bankers, lawyers, and even the mayor were there to help in whatever way they could. Though there isn't much a twelve-year-old can do at a construction site, I was kept busy by being a gopher and supplying nametags. My mom was at the house every day that summer. She put her heart and soul into it. After a long, hot summer and a lot of forfeited Saturdays of sleeping in, the moving day finally arrived.

September 25, 1994, was my thirteenth birthday as well as the first night I was able to sleep in a house where the roof didn't leak every time it rained, the heating worked properly, and the walls didn't show forty years of different colored paint.

It's been five years since my building started with Habitat for Humanity and although the roof is over my head, the work will never be complete. Bricks and mortar only go so far, but love and care have no end.

No More Gunfights

In December 1995 I was privileged to attend the ten-year celebration of South Brevard Habitat for Humanity in Melbourne, Florida. Linda and I were invited by Kim Gabriel, the executive director of that outstanding affiliate. One of the events of the celebration was dedicating the new Habitat home of Silas and Pamela Scott and their four children.

Future Habitat homeowner children in the Philippines.

Linda and I presented the keys to the family for their new Habitat house. They had been living in a tiny two-bedroom apartment, with all four children sleeping in one of the bedrooms. Their former house was in a really rough neighborhood with frequent fist- and gunfights. Pamela said that living there was "like a constant battle, with our trying to teach Christian values and 'reality' doing a better job at undoing them."

Kim Gabriel reports that the children are doing well in their new house and neighborhood. Their grades have improved in school, and the whole family is thriving.

Showing Off His House

Betty Wayland of Rapidan Habitat for Humanity in Locust Grove, Virginia, shared a truly heartwarming account of the difference their first house had on a ten-year-old boy. The family, consisting of a single mother and four children, lived in a trailer with two bedrooms and no running water or bathroom facilities. The only electricity was a cord running from the house next-door. An electric heater provided warmth. For eighteen years the family had lived in that trailer.

When the family moved into their new Habitat house, they had to learn many things about fuse boxes, fans in the bathroom, and so forth. The ten-year-old son was so enthused about the spray hose on the kitchen sink that he volunteered to do the dishes. And months later he was still at it! Betty says that the young man is proud to show his classmates his new home. His grades have also improved, and he is active on the basketball team.

Blended Family Has Space to Blend Right

Janet Deckert, president of Habitat for Humanity of Martin County, Florida, tells a similar story about the children of a family that moved into their new Habitat house the week of Labor Day in 1997. A blended family of nine, brought together by death and a desertion, were living in a house with two rooms. The children piled on the floor to sleep at night.

Now, though, the family is well ensconced in a roomy Habitat house, and Janet reports that they are "one happy family." One of the volunteers donated a computer to the children to help with their schoolwork. Their grades have improved, and one of the boys was on the football team.

First a Hammer, Then a Scholarship

Patricia Curtis is a wonderful living example of the "more" in *More Than Houses*. This outstanding young woman was raised in a shack in Harlem Heights in Fort Myers, Florida. She was a good student even in the years when she and her family lived in the bad house. But she blossomed even more when the family moved into their Habitat house. Susie Ellis, development director of Habitat for Humanity of Lee County, asked Patricia's mother, Rosalee, "Do you think Habitat helped her?"

"Oh, yes!" Rosalee exclaimed. "She first learned to use a hammer at Habitat. And she learned compassion, cooperation, and thankfulness!"

Rosalee told of Patricia's change in spirit after the family moved into their new Habitat house during her junior year in high school. She said Patricia seemed to blossom and was finally able to bring her friends home with her. Rosalee said she understood Patricia's excitement. "I didn't sleep for at least two weeks after we moved in because I couldn't believe it was mine! I would get up in the night when everyone was sleeping and just walk around and look at it. I still love the house!"

When this book went to press, Patricia was majoring in nursing at Georgetown University in Washington, D.C., on a Georgetown Leadership in Nursing scholarship.

Without a doubt, a decent place to live has a tremendously positive effect on children. That alone would make the work of Habitat for Humanity worthwhile. The impact moves outward from there, like a wonderful ripple effect to the whole family.

CHAPTER 3

Transforming Families

Habitat for Humanity is building much more than houses.
By building hope it is building relationships, strengthening
communities, and nurturing families.
—PAUL NEWMAN, ACTOR, HABITAT HOME BUILDER

Au aspects of the Habitat program are structured to nurture families and break the poverty cycle—not just provide an affordable house," concludes the AREA study. And it's true. Habitat works with and for the families it moves into each home to help them succeed. How?

Usually before homeowners move into their homes, they have gone through the affiliate's careful selection process and received training in a wide variety of areas, from home maintenance and budget preparation to gardening and parenting. Habitat for Humanity International affiliates build pride and confidence by involving families in the construction of their own homes and encouraging them to help other families become Habitat homeowners. After the move-in, affiliates often provide ongoing support in the form of low-cost maintenance, repairs, fixtures, furnishings, and the like.

The AREA study also noted: "As a model for efforts to provide home-ownership opportunities for very-low-income families, the Habitat program appears to work very well. By providing not only a one-time, up-front subsidy (no-profit, no-interest mortgage), but also ongoing nurturing to overcome new financial hurdles as they arise, Habitat makes housing ownership possible for those for whom even small unexpected expenses can cause financial crises."

According to the report, Habitat home buyers have benefited from homeownership in "monetary terms" and in terms of "qualitative lifestyle changes."

Incomes Have Increased

Monetarily, Habitat homeowners pay, on the average, only 24 percent of their incomes for total housing costs—including mortgage payments, real estate taxes, homeowners' insurance, and some maintenance. Only 10 percent of their income is for mortgages.

That alone is interesting and far below the norm, but this is the fascinating part. The report found that, for many homeowners, the percentage of income devoted to the costs of housing actually decreases after the purchase of the Habitat home—because the homeowners' incomes have gone up. Most families that participated in the survey had previously lived in rental housing. In most cases, monthly rental costs were about the same or sometimes more than their current total housing costs as Habitat homeowners. Former renters, especially those who lived in subsidized housing, felt better able to control their finances. They believed that their housing costs would be constant, and as their incomes increased they would not have to make higher house payments.

Lifestyle Quality Has Increased

The AREA study, after asking hundreds of owners the same questions, found this to be true:

The most frequently mentioned benefit of homeownership was the pride and increased stability that the family got from feeling safe and secure about their home. Most homeowners interviewed had no plans to move up or capture appreciation. They planned to keep their homes forever and saw them as something to pass on to their children.

Since homeowners had received better quality, more spacious houses, they were enjoying the ability to entertain company without embarrassment. The majority of interviewees could not name any specific positive effects that homeownership had on their employment opportunities; however, those who claimed a positive impact on their jobs explained consistently that the reliability of constant home payments had given them new flexibility to plan for their future, return to school, look for a better job, etc. The burdens of home-ownership were perceived as few.

I can almost hear every family interviewed. The impact of a decent place to live on a family is so profound yet so universal. The study's findings only underscored what I have heard and experienced since the Johnsons moved into that first home at Koinonia Farm and transformed their home life and their future.

The following Habitat families have invited us into their homes. Their stories are eloquent confirmations of the AREA study's findings concerning the powerful transformations brought about by the Habitat houses and all of the faith and love that go along with them.

Future Habitat homeowners in Coatetelco, Mexico, express joy as their lot of land is quickly being turned into a home. Photo by Robert Baker.

A Piece of the American Dream

Mitchell, Maggie, and their children were living in public housing when Mitchell's employer spoke to him about Habitat for Humanity and suggested that the family apply for a Habitat house. They did so and were approved.

Joe Abernethy, executive director of Habitat for Humanity of High Point, North Carolina, said the affiliate's volunteers partnered with Volvo to build the house. When it was time to start the framing, each child wrote his or her name on a stud in the wall of their bedrooms. All the children helped build the house. One daughter wants to pursue a career in medicine. A son is on an award-winning high school debate team and has attained Life rank in Boy Scouts. Another daughter joins her father in live and recorded sessions on radio, television, and personal appearances in the community.

Mitchell expressed it best: "I am thankful for a stable spot to raise my family. We are living a piece of the American dream. The children are motivated to excel. Truly, Habitat builds lives, changes communities, and gives witness to the power of love."

Set Up to Succeed

Antoinette Hingleton, a divorced mother of four children, works as a clerk for a local insurance company. Her small, crowded apartment was full of people, belongings, and furnishings. When a cousin told her about the houses Habitat for Humanity was building, Antoinette decided to apply. Her application was approved, and Wesley Memorial Methodist Church became the sponsor for the house.

Antoinette has very pleasant memories of building her house. She had fun choosing her cabinets, shutters, and siding. Everyone worked well together, helping each other and bringing the neighborhood into closer harmony. Then the happy day came. Two days before her birthday, Antoinette's house was dedicated, and she received the keys. The whole family participated in the ceremony and beamed with joy.

All of Antoinette's children have blossomed in their new home. Danforth is a martial-arts expert, plays football and basketball, and is a honor-roll student and citizen of the year at his school. Justin plays piano; takes part in art, drama, and music activities; serves as secretary of the student government; and was named student of the year in his middle school. Tirrell is an outstanding football player. He also loves to draw and play saxophone, and he is a deep-thinking computer whiz. Nicole is at the University of North Carolina at Chapel Hill majoring in accounting and advertising. In high school, she had an outstanding academic record and was a member of the National Honor Society and several other student organizations. She served as a volunteer with the human relations commission, Carson Stout Homes, Habitat for Humanity, and was president of the Beavers 4-H Club for two years. She hosted *Teen Talk* on radio station WMFR for a year.

Antoinette described her experience as a Habitat homeowner: "Habitat is set up to help you succeed. It is great for people who want to improve their situation but have no way to do it because of limited finances. Helping me to provide a home for my children has made Habitat very special to me. I will gladly partner and help in any way I can." She went on to affirm the home's impact on her children. "Being in our own home gave the children more self-esteem. They have more space, positive concepts, and determination to achieve. I am glad that I was able to get some help that enabled me to be a full-time mother and provide for my children."

New People, New Perspectives, New Hopes, New Dreams

Claudia Doodick of Salisbury, Connecticut, wrote a powerful letter about the effect of Habitat for Humanity on her family: "I used to go to bed each night and pray to God to help me find a way to get out of the apartment complex in which I was living. We had a small, two-bedroom duplex sandwiched between two other apartments. We had no yard. Fights occurred regularly among tenants in the other fifteen apartments in the complex. My car was damaged. My daughter was harassed and assaulted. She would no longer go outside.

"My son was seriously injured in a car accident. After three months of hospitalization and in-patient therapy, he came home in a wheelchair. He slept in a hospital bed in the living room. We were really cramped. None of us had any privacy. My daughter retreated to her room. My son often spoke of being in the way. It was sad to watch what was happening to my family. It became more urgent to find a bigger, more comfortable, safer place to live. I continued to pray my nightly request to God. Then I heard of Habitat for Humanity and how a house was going to be built in our town. I filled out an application. I will never forget the day I was notified that I had been selected to be the future owner of the house to be built in Salisbury, nor will I forget the following months of waiting for the house to be completed. All through that time, my children and I were introduced to many new things: new people, new ideas, new perspectives, new hopes, and new dreams. I saw my children's feelings change from despair and fear to hope and confidence that the days ahead would be better."

As for herself, Claudia manages a local convenience store—gas station and sees lots of local people every day. "As word spread that my family had been selected, people wished us well, told me they were happy for us, and offered to help with the house in various ways," she wrote. "My self-esteem and feelings of worth skyrocketed. I found that people really liked me. I felt a new sense of belonging in the community."

Hope is a great energizer, and the anticipation of Doodick family's moving into a new home changed her son's attitude and his health. "My son slowly graduated from his wheelchair to a walker," she wrote. "He became determined to use a cane so he could walk to the building site which was a thousand yards from our apartment. He was determined to help put some of the required four hundred hours of sweat equity into the property, and he succeeded."

The day the exterior walls were raised on the second floor, her entire family was present: children, sisters, nieces, nephew, father, and stepfather. "Although we were hard at work, we and all the Habitat volunteers, it was like a party. Being part of such a diverse group of people gave me the feeling that we were part of the entire universe. That feeling sent chills running through my body. When our home was completed and the well-attended dedication ceremony was held, I looked out at all the people. Some of them had become very dear to me. There were family members, coworkers, neighbors, old friends, new friends, and future friends. God's presence was everywhere and every person looked like an angel to me.

"This is what I told them all that day: 'You are all angels sent from God to be part of our extended family. I feel that way deep in my heart. God does work through people, and I haven't met any people more loving, devoted, generous, and kind than the people I have met through Habitat for Humanity.'

"Now that we are living in this wonderful home, my daughter no longer hides in her room. She doesn't hesitate to invite her friends to come and visit her. She has a newfound confidence. She comes home from school every day whereas she used to look for other things to do in order to avoid coming home to our old apartment where she was so unhappy.

"There is so much more I could tell you about the changes we have experienced due to our involvement with Habitat for Humanity, but it would become a book in itself for me to do so," she wrote. "So I will end here by sharing with you a prayer I say every morning to help me focus on the kind of person I want to be and the kind of life I want to live:

"'Lord, stay with me throughout this day. Help me to see with Your eyes, hear with Your ears, speak with Your words, think with Your understanding, love with Christ in my heart, walk on Your path, and serve with Your hands. In Jesus' name I pray. Amen.'"

A Home for Holly

Tom Bridge, past president of Habitat for Humanity of McDonough County, Macomb, Illinois, tells an exciting story of transformation in the family of Perry and Nancy Sinnett and especially of their daughter, Holly. She was born with hydrocephalus, which is water on the brain. Holly has a shunt from her brain to her stomach to release the fluid. She also has a seizure disorder, cerebral palsy, and wears special glasses for an eye problem.

Holly was unable to get around in their old house, which had only one bedroom. She had to crawl or her father had to carry her from room to room. The bathroom was not accessible to her, so someone always had to help her use it. The Sinnett family moved into their new Habitat house in December 1997. Their house was the first built by the local affiliate. Here is Tom's account of the transformation:

"I met Holly and her family on a cold winter day when I went with the Family Nurturing Committee to visit the Sinnett family. We were discussing floor plans for their proposed Habitat home. When we arrived, Holly was playing on the living room floor. She was very shy. We moved to the dining room to spread out the plans on the dining room table. Holly's dad brought her into the room so she could be a part of the meeting.

"During the days of construction, Holly was almost always present," he remembered. "Her parents would bring her over so she could watch us build. I remember clearly the day we finished the subfloor. Her dad lifted Holly onto it and took her to where her new room would be. She remembered that spot throughout the entire construction period. If anyone asked her, she could take them to her room regardless of the stage of construction. At the dedication of the completed house, I heard Holly exclaim to many visitors, 'Come and see my new room' or 'Come and see my new bathroom.' I watched as she rolled her wheelchair into the bathroom and showed everyone how she could now wash her hands without help. On that dedication day, no one had more enthusiasm than Holly. She was so proud of her new house and of what she could now do. She told everyone how her bedroom was going to be arranged and how it would be decorated.

"I attended Holly's fourteenth birthday party about eleven months after they moved into their new home," he went on. "Immediately, I could tell how the new home had made a big difference in Holly's life. Her room was decorated with lots of stuff from the movie *101 Dalmatians* and was well organized. Holly can now get around her home in two ways. Since the new home was made accessible for her wheelchair, she can go everywhere. But now she can do even more. Since moving into the home, she has been able to practice and get around with a walker. The wheelchair was impossible before and just to have the use of it at home was a major accomplishment, but now add to that the ability to use a walker. It is truly amazing. Holly now has mobility never possible or even imagined in the previous home."

Holly's mother also mentioned that this year they had all been a lot healthier. In fact, she said that by now one of them would normally have been very sick with a cold or flu. "But this new and much warmer home was keeping them all a lot healthier," Tom added. "My wife and I went to [Holly's] school Christmas play a few weeks ago, and after the play Holly asked us to go to her classroom and see her schoolwork. Well, she took off down the hall in her walker faster than any of us. A year ago, she could not do this." he said. "It is a good feeling to see a new Habitat home protecting a family, but no feeling is as great as the one of seeing a child grow beyond prior expectations in that new Habitat home. That once very shy young girl is now a growing young lady both physically and mentally. She now talks to me, and every time we meet she gives me one of Holly's hugs."

Pride in Ourselves—"We Don't Just Exist, We Live!"

In Pitman, New Jersey, the Chews—Glenn Sr., Diana, Dottie, Jessica, and Glen Jr.—were living in a small apartment below ground level with noisy, inconsiderate neighbors. The rent was $650, which was a tremendous burden on the family. Another big problem concerned the health of their daughter Dottie.

"Dottie is multiple disabled," Diana explained. "She is in a wheelchair, and I used to have to carry her wheelchair into the house and then carry her, with our two smaller children crawling behind me. Due to my daughter's medical condition and the way we were living, we were always stressed, fighting, and unhappy. I think we felt like this was all there was, a vicious cycle we'd never stop. Then we went through the application process for a Habitat house. We met individuals from Habitat who were so nice to us. They treated us like we were special!

"We now have a beautiful three-bedroom house that has wide hallways and doors for Dottie's wheelchair. They put in a ramp for us and gave us a yard to let our little ones play in. This all changed our lives," she said. "The children are happier; we have pride in ourselves again and have real lives. I have gotten involved with my children at school. In fact, I have taken on the presidency of the parents' group association at Dottie's school. Habitat has made us a better family. We are able to be a part of a community and give something back to our community. And our new house has enabled us to be a real family. We don't just exist; we live!"

Beverly Currier, chairperson of the nurturing committee of their affiliate, Gloucester County Habitat for Humanity, wrote this about the remarkable Chew family: "The Chews have been Habitat partners in the very best sense of the term. They go far beyond their mortgage obligation. Glenn has used time off from his factory job to help deliver materials to work sites. Diana has written thank-you notes and spoken at general meetings. She has been room mother for her younger children and has taught classes at their church. I am sure Habitat has made a dramatic difference in their lives even though they were already a fine young family. The Habitat experience has freed them to become an example to other families and real community helpers."

In July 1998 Linda and Millard Fuller spent the night with Habitat homeowners Bernard and Juliana Gabriel in Hatton, Sri Lanka.

A New Me

Teresa Burton of Manchester, Tennessee, boldly states that the Habitat house she and her family received from Highland Rim Habitat for Humanity brought about the biggest change ever in her life. She says she received not only a house but also "love, self-esteem, guidance, and a new extended family. The love I've received is such as I've never seen anywhere else. I was never judged on what I had, what I didn't have, or what I could give. I was accepted for me: a child of God in need. Habitat continues to improve my life on a daily basis with different programs, such as budgeting classes. Habitat has never forgotten me, and I will never forget them. I believe in this program so much that I continue to work on committees, help with new families, and do public speaking, and I am currently helping organize a homeowners association in our area. The whole idea of this program is to give a hand up, not a handout. So I continue to give back the things I was given. I didn't just receive a clean, healthy, and beautiful home; I received a new me! That was, and is, as important as the home itself. I hope and pray every day that I can touch one person the way Habitat has touched me."

New Heights of Accomplishment

Overcoming obstacles and rising to new heights of accomplishment seem to be hallmarks of so many Habitat families. Two stunning examples of this are from Habitat for Humanity of Wayne County, New York (Macedon). Irene Fadden of that affiliate tells about the Bowman and Pratt families: "Ed and Kristy Bowman and their three children were selected to be the homeowners for the second home built by HFH of Wayne County. The house was started, but unfortunately, the affiliate ran out of money. The house sat empty, unable to be finished. At that time, the Bowmans lived in a small two-bedroom trailer. Ed suffered from severe depression and was unable to work. The family was living on Kristy's income as a home-care aide for the elderly. During the construction of the house, Ed put in his sweat equity hours when no one else was working on the house. In February 1992, their home was finally completed and the changes began. Ed was able to get a part-time job as a car salesman and then at a grocery store. Then he got a full-time job with the Wayne County Behavioral Health Network as an Activity Aide," Irene said. "He is also featured on Sundays on a local radio station, playing the Oldies on Memory Lane. The children have thrived in their home. Jennifer is an honor student. David was in a self-contained special education class but is now able to be in the mainstream class. Eddie is also doing well in school.

"The Pratts—Vern, Sandy, and four children—were living in two bedrooms with Vern's parents," Irene went on. "When selected for a Habitat house, Sandy knew she was dying from cancer and she wanted a home for her family. Construction got underway in May 1992, and it took three months for the house to be completed. Sandy lived in the home for eleven months before her death. Vern was able to pay off the medical bills and other bills that mounted during Sandy's illness. This was no easy feat, but the support of Habitat through nurturing and budget counseling made it possible. Vern has since remarried. He and his three youngest children bless Habitat gatherings by sharing their gift of music, and Vern continues to be an advocate for Habitat in the community."

All across the land, transformations are happening. Not every story is a story of triumph, but the positive experiences far outweigh the negative. The names of the people change and the towns and cities are different, but the stories are the same. Old dilapidated houses or apartments change to modest but good, solid houses. Renters become homeowners. Despondency and a sense of hopelessness give way to rejoicing and buoyant hope. Instability in families is transformed into rock-solid stability. Poor performance or failure in work or school change to promotions and success. All because these good people have found a simple, decent place to stand, which then gives them the chance to move forward.

Okay, We Are Making It

In the West, in the city of Seattle, the Amaral family was living in a run-down old house. They applied for a Habitat house and were accepted. The entire family worked on the new house as it was going up. They labored before and after work, on weekends and on holidays. It took two years to complete. The house is modest—three bedrooms and only 1,050 square feet—but tall ceilings make it seem larger. On dedication day, the family members were given an opportunity to express their feelings about the new house. When it came Ana's (the mother's) turn, she couldn't speak. She just stood before the microphone and wept.

Their monthly mortgage payment is about half what they were paying in rent for the old house.

Their son, Roberto, graduated from high school and enrolled in the local community college. Ana exclaimed, "Before this house, I always thought maybe my kids wouldn't be able to have an education, because all we do is pay rent and bills, and I thought we wouldn't be able to do anything. After the house, everything changed. We can save some money for the kids. My son is going to college now, and I think, *Okay, we are making it.*"

House-Proud

Another story of triumph from Washington State comes from Ellensburg. Joyce Nelson of the Ellensburg Area Habitat affiliate was a neighbor to a woman named Margaret. After Margaret moved away, Joyce got word that she was not doing well. Margaret had lost her house to the bank. Due to illnesses, Margaret and her children were forced to move in with her parents.

Margaret filled out an application for a Habitat house and was approved. As the house went up, Joyce stayed in touch with her former neighbor and enjoyed watching Margaret's self-esteem grow. She got more hours of work at the local food-processing plant where she worked, and she earned seniority status. Overall, things really looked up for Margaret and her children when they moved into their Habitat house. Margaret is now "house-proud," as they say in the West. "And," Joyce concluded, "it looks real good on her, too."

Living in the Church Basement

A long, hard struggle led to homeownership for the Gillis family in Monmouth, Oregon. In 1993, the house they were renting was sold, and the family of six had to move. With only one meager income from the father, the family could not afford anything big enough for them. They ended up homeless. For two months they lived in the basement of a church. Then a friend of the church pastor rented a house to the family. They learned about Habitat for Humanity and applied for a house. It took three years for construction to begin. In 1997, the family moved in, and finally, on April 15, 1998, the sale was completed, making the Gillis family one happy bunch of people.

Sheila, the mother, expressed her feelings about their dream house. "Every morning I wake up and feel so blessed to be in such a beautiful home," she said. "It is our first step to security. The payments are low and affordable. The home is clean and functioning. We are in a community that is growing dramatically. The children are having an opportunity to really experience what it is like to belong to a community. This will be their town when they grow up. They will take pride in it and continue to make it a better place to live for their children."

It Makes Me Whole!—Why Habitat Succeeds
over Most Social Programs

Sally Marchington, a Habitat homeowner in Medford, Oregon, wrote movingly about how Habitat and God have made her whole. She said:

Habitat for Humanity–Rogue Valley and God have made me whole. The hand up I needed was extended when I was teetering on the fine line between hope and hopelessness. Now I extend my hand to others as a Habitat for Humanity–Rogue Valley volunteer. The strong sense of community, family, and knowing I am making the world a better place are very fulfilling. It makes me whole! The sense of peace a single parent can seldom experience, comes with this contribution. People are helped to help themselves. This builds strong self-confidence and dignity (the building blocks for success).

This is why Habitat for Humanity is successful where other social programs fail. Put-downs, stereotyping, criticism and mental abuse lead to failure, giving up, and social dependency. Supportive, educational, caring programs that require responsibility (like Habitat for Humanity) strengthen self-confidence and families and create a winning situation for all. The family, Habitat for Humanity, the community, the nation, and the world are all stronger in the end. Throughout the process, God is smiling down on us all. May God continue to bless our Habitat for Humanity members worldwide.

Amazingly Similar Worldwide

Speaking of Habitat members worldwide, there is now Habitat work in more than sixty nations. And the stories of the transformation of individuals and families are amazingly similar all around the world.

In January 1998, Linda and I along with our daughter, Faith, and Habitat's director for Africa and the Middle East, Harry Goodall, toured our work in three African countries. In Zambia, we met Sylvester Tembo. He did quite a bit of translation for me as I spoke at various locations.

Sylvester is a Habitat homeowner. In 1993, when he was selected to have a Habitat house, he did not have the money to hire a bricklayer to put up the blocks for his house. The local affiliate offered to train him while his house was under construction.

He built the toilet first then laid all the blocks for his own house. He loved the work enormously and discovered a talent he never knew he had. He also loved his house. Sylvester exclaimed after finishing it, "I like it so much because it came from my own hands. Habitat has not only made it possible for my family to have a house, but it gave me a skill that will go a long way to support me and my family in the future."

On dedication day, Sylvester's house was chosen as a model Habitat house in the area. Since then, Sylvester has been hired by many people to build houses for them, and that work has earned him a substantial income with which to sustain his family.

In northwest Tanzania, Habitat for Humanity has built more than nine hundred houses. We visited that work also in January 1998, and we were told this story about the effect a Habitat House had on cleanliness and health in the one of the very first families to own a Habitat home built by the Korogwe affiliate in 1997.

Louchia Rosie Mkande had previously been living with her six children in two rented rooms. The place was congested and there was no place for the children to play. The poor living situation resulted in frequent illnesses, such as coughs, colds, and malaria. Dedication day in February was a very special time for her and the other four families because the president of Tanzania, Benjamin Mkapa, was there to hand her the keys to her new home. In his remarks on that occasion, he said that the families should be grateful to God and to Habitat for Humanity for their new homes, and he strongly encouraged each homeowner to faithfully make their house payments so other families could also get houses.

When it was Louchia's turn to speak, she said, "I do thank God for this beautiful house. Now my children can play freely, have enough space to study and have privacy and no more bother from a landlord. I will be able to maintain cleanliness as I wish."

Personal Burdens Lighter

Christine Makokha became a Habitat homeowner through the Katira affiliate in western Kenya in 1996. She says that her involvement with Habitat for Humanity has changed her life and her children's lives tremendously. She says her new house made her feel great and gave her peace of mind and love for people whom she never thought could be her friends earlier. "As I continue working with Habitat and get more involved in helping others who long for love and shelter, I find that my personal burden gets lighter," she said. "This is because I don't keep thinking of my personal problems but instead think of a neighbor who doesn't have shelter and who feels unwanted in the community.

"Habitat has helped me learn more about faith and myself. It has enabled me to develop new skills in construction so that I can now educate homeowners-to-be on building construction without assistance," she continued. "Habitat has transformed me in that neighbors no longer overlook me as a widow. My children feel loved and wanted in the community."

Other Cultures, Same Helping Hand

Another homeowner in that same affiliate in western Kenya experienced a transformation that may sound strange to a Western reader but reflects reality in that part of the world. Here is her story:

"I lived in a grass-thatched house since 1967 when I got married. The house started leaking after a few years. When it rained, I had to move bedding, furniture, and everything to one place and cover them with polythene paper to save them from getting wet. My children were frequently sick with malaria and pneumonia. I could not cook because the fireplace would be filled with rainwater. When I was not at home and the rain started falling, I was worried because that day we would not have a good sleep due to the wet beddings. The house was not even good to look at. And when there was moonlight at night, the inside of my house was like daytime because the light came through the roof.

"All the above made my husband desert me," she said. "He got married to a second wife for whom he built a good house. He forgot all about me, and when visitors came he would not bring them to my house. This contributed to my hating my co-wife very much plus the husband himself. There was war every time I met with my husband just because of the state of the housing in which I was living."

At that sad time in this woman's life, she encountered Habitat for Humanity. She applied for a house and was accepted. It was built over a period of a few months, and she moved in amidst great rejoicing. She wrote movingly about what happened to her as a result of getting the new house. "There is one thing to which I must testify: Habitat for Humanity does not only build houses but also transforms people, and this I am saying from my innermost feelings," she said.

"I am one of the people whom Habitat was looking for and found when I had lost hope of living on this earth. Since I received my Habitat house in August 1996, miracles have happened in my life and now I am a reformed woman: (1) My husband has come back to me thus our marriage renewed; (2) My children are no longer sick; (3) I am no longer worried about sleeping on wet bedding; (4) The understanding between the second wife and me has been reformed; (5) I have met and made friends with many people from other countries that I never thought I would meet in my lifetime, especially 'wazungus' (foreigners)."

Habitat homeowners the world over not only report great changes in their lives, but they express a strong desire to give back to help others as they have been helped. Indeed, this aspect of the ministry is one of the major reasons for Habitat's success. Every homeowner is a partner, and they each receive a blessing, but giving back is an even greater blessing.

Giving Back

Sherri Moore of Statesville, North Carolina, described how she and her husband, Roy, want to give back. Their family was living in a very cramped two-bedroom apartment. Also, their son, Joshua, had a mild form of cerebral palsy and was extremely active. This made it impossible for Sherri to work regularly, hence their income was limited to that of her husband.

They learned about Habitat and applied for a house to be built by Irdell-Statesville Habitat for Humanity. The family moved into their new Habitat home and the transformation began. Roy now has a business that is prospering. The children are flourishing. Joshua is exceeding everything the doctors predicted. Sherri exclaimed, "God has shown us that one day our business will build a Habitat house."

The Woodward family of Dublin, Georgia, is another family that wants to give back. They moved into the second Habitat house in Laurens County on Christmas Eve in 1993.

The family has blossomed since moving into their new home. Ed got a responsible job helping a businessman with his service station and farm. Teresa received her GED and has plans to become a CPA. She and Ed have ambitions for their four children, too, desiring that they be well educated and "go off and be something wonderful."

In building their house, Teresa and Ed fulfilled the five hundred-hour sweat-equity requirement. After that, they quit counting but kept on helping out on subsequent houses. They call the volunteer work their "Saturday habit."

Teresa and Ed have a dream. When they finish paying back their house loan, they want to make another family's dream come true. "We want to one day give a Habitat home," they said.

They have had the joy of receiving. Now they want to expand their joy of giving by donating the money for another house.

Dia Hadley, executive director of Spokane, Washington, Habitat for

Humanity told a similar wonderful story of a homeowner giving back. The affiliate had an urgent need for thirteen thousand dollars to pay some bills. Dia had no idea where to get the money. Then, out of the blue, a check in that amount came in from a homeowner who had received a settlement in a case. Dia said the money was like a blessing from heaven and doubly so because the payment was not required and was given because the home-owner wanted to help others as her family had been helped.

Giving back is often in the form of personal involvement with the ongoing work of Habitat for Humanity. Karen Roach is an excellent model of this kind of involvement. She and her daughter were chosen to have a Habitat house built by Adams County Habitat for Humanity in Gettysburg, Pennsylvania. They moved into their new home in early 1996. When asked about her Habitat experience, Karen replied, "I give God the glory for what has happened. I know God watches over me. I am awed by it." Karen serves on the family selection committee of the local affiliate and is a proud spokesperson for the fine work being done by Adams County Habitat.

Uptown Success Story

Rosie Simmons of Chicago is another person who has not only been transformed by her Habitat experience, but who gave back in numerous ways. She was featured in the November 1997 issue of *Ebony* magazine, and her story is a gripping one. Rosie was living in a miserable apartment in uptown Chicago, and her family was in disarray. Then she was chosen to have a Habitat house. Her life began to turn around. She got a job with a bank in downtown Chicago. Over the next four years she was promoted four times. She now holds a very responsible job at the bank, and she works closely with Uptown Habitat for Humanity, the affiliate that built her house. Among other things, she is a frequent and extremely eloquent speaker for Habitat at churches, civic clubs, and other organizations.

Elaine Warner French, who lives in a Habitat house in Hanover, Pennsylvania, with her daughter (who calls their house a castle), stated that their house was a dream come true. She claimed that she would be a lifelong volunteer for Habitat.

Rocky Road to Her Castle

Ellender Rhodes of Columbia, South Carolina, called her new Habitat house a castle. The road to that castle, though, was a long and tortuous one.

Ellender was raised by an alcoholic grandmother and suffered a great deal of verbal abuse and neglect. By the time she was thirteen years old, Ellender was hanging around bars. At age fourteen she dropped out of school, and at fifteen, she gave birth to her first child. During that same year Ellender downed a full bottle of aspirin and headed to the train tracks intending to kill herself. Luckily, someone stopped her.

At age eighteen, Ellender became a mother for the second time. Her drinking worsened. In 1990, she was sentenced to a women's correctional center where she received help for her addiction.

After her discharge, she lapsed and was admitted to the state treatment center. At that facility, Ellender began to listen. Upon her release, she entered a women's shelter in Columbia. There she received dental care, worked on her social skills, and found a job in a local factory where she was promoted four times in one year.

In February 1992, Ellender entered Trinity Housing Corporation, which provides homeless families housing for up to two years along with counseling and training in job skills, budgeting, family skills, and conflict management. While at Trinity, Ellender regained custody of her youngest daughter. She also worked as a waitress at night and attended school during the day to work toward her GED.

In 1993, Ellender learned about Habitat for Humanity. She applied and was accepted for a house. A year later she received a Bible and the key to her new house. "Well, as you all know, I have the key to my castle," she said on that happy occasion. "I am so grateful I can't put it into words." Four years later, Ellender was still living in her castle with both daughters and faithfully making her mortgage payments every month.

A Habitat Home and a Habitat Job

Judith Woods of Charlotte, North Carolina, got not only a Habitat house, but a Habitat job. "It was such a blessing that Habitat was there for me in my time of need for a house," she wrote. "Now I am employed by Habitat, and this is the best place that I have ever worked in my life."

Lexington Area Habitat for Humanity in Lexington, North Carolina, called Alice Hamilton their "shining star." Alice, her husband, their two children, and a nephew applied for and received a Habitat house. After the family moved in, she became a truly incredible community leader.

Alice was a founding member of the Lexington Housing Centers for Disease Control. She became a member of the board of Davidson County

Community Action Corporation. Then she was elected chairperson. She also serves the Lexington Area Habitat for Humanity as chairman of the nurturing committee. She completed the year-long Davidson County leadership course through the local community college. Finally, Alice was president of the Junior High and Senior High School Band Boosters Club.

Before moving into their Habitat house, Alice had never been a volunteer for anything. Lou Adkins of Lexington Area Habitat exclaimed, "It has been so marvelous for everyone on our Habitat Board to watch Alice grow and stretch her wings."

Self-Sufficiency

Chinita Gunnings of Gastonia, North Carolina, was unable to speak or hear. She claimed that it would have been impossible for her to obtain a home of her own without Habitat's help. She proclaimed that the house "has changed my life by giving me the chance to become independent and self-sufficient. I now have the security and peace of mind to know that my child can play in her own fenced-in yard. I now live in a neighborhood with neighbors who care and have the same goals as myself."

Embraced by a New Family

Countless Habitat homeowners talk about the love they receive from their Habitat partners. They feel like they have been embraced by a larger family of people who really care for them. Karen Zerphey of York, Pennsylvania, shared the following:

"Shortly after moving into my home, I fell and broke both of my ankles and ended up in a wheelchair for the next four months. When Habitat announced my accident at the board meeting, the different religious leaders went back to their churches and asked for prayers and help for the boys and myself. Total strangers came to my house with food. Offers to clean for me, to bathe the boys, to run errands, etc., came flooding in. Never in my life have I seen or felt such warmth from people who reached out to help, not because I was from their church, but because we are all children of God and because of that we all are an important part of this life. Habitat gave me a chance to own a house, but, along with it, I received a family and I received a new life."

Ruth Ann Stroble and her preteen daughter, Shannon, were living in an unheated attic apartment above an older daughter's home when they learned about Habitat for Humanity. She applied to Clark County Community Habitat in Springfield, Ohio, for a house. At the time, Ruth Ann was on public assistance and, by her own admission, "felt worthless." But Ruth Ann said that Habitat for Humanity changed all that. She believed that God used Habitat to show her the way to a better life. Since living in their new Habitat house, Ruth Ann has graduated from college with an associates degree in social services. She secured employment as a family service worker at London Head Start.

April and Brad Morgan and two of their four children moved into their Habitat house in Earlville, New York, in 1995. The house was built by Hamilton Area Habitat for Humanity. Since moving in, the family has become quite self-sufficient, raising beans, squash, tomatoes, and other vegetables in a quarter-acre garden. They also have been raising calves, pigs, and a goat. In January 1998, April enrolled at the State University of New York at Morrisville. She said that the only reason for entering college, after years of wanting to do so, was because she now had the stability of a permanent home.

In 1991, Mary Ellen Stubb, her husband, Joe, and their five children were living on Joe's income of $6.50 an hour in a very inadequate house. A home of their own was a distant dream they thought they'd never realize. Then one day a flyer from school told about Habitat for Humanity. The family applied for a house and was chosen to receive the first one built by Habitat for Humanity of Missoula, Montana.

The family moved into their new home on the day before Thanksgiving in 1991. Mary Ellen said they finished moving at 2 A.M. on Thanksgiving morning. She said they cooked turkey and ate on top of boxes. "It was the best Thanksgiving ever," Mary Ellen exclaimed. In the years since, the family has remained involved with Habitat, which built fourteen more houses by the end of 1998. Mary Ellen has served on the local Habitat board and was for a time chairperson of the family selection committee. She has also supported Habitat's inclusion in Missoula County's charitable-giving campaign.

After moving into their Habitat house, Mary Ellen went back to school. She graduated from a vocational-technical school in 1994 with a degree in accounting. After a couple of interim jobs, she went to work full time in October 1995 in the county accounting office.

A very similar story to that of Mary Ellen Stubb is that of the Martinez family in Oklahoma City, Oklahoma. In May 1993, Nancy Martinez, husband José, and their five children moved into their new Habitat house from a small two-bedroom rental house, which, as Nancy said, had an ample supply of mice and roaches, bad plumbing, and only one small heater, which was in the living room.

Nancy said that moving into their new house changed their lives. One of the biggest changes was the opportunity it afforded her to return to school and finish her degree. She had to drop out of college in 1992 because of the high cost of housing. Those expenses were dramatically reduced after the move. Their house payment was less than their rent, and their utility bills were also lower. With the savings, Nancy reenrolled in college. She finished her degree in alcohol and drug counseling in 1997 and took a job with a community agency. In that position, Nancy was able to spread the word about Habitat for Humanity to others who needed a good place to live.

Cycle of Welfare Broken

Barbara Perry of Boonville, Missouri, got a job because she was chosen to have a Habitat house and because of the confidence it gave her. Ken Askren, president of Boonslick Habitat for Humanity, told the exciting story of this fine and courageous woman. "The issue before our board was reaching out to a family in need who really was walking on the edge of ability to fully partner with Habitat," he said.

"We prayed and we heard the Lord saying to us, 'Do what is right. I will show you the way!' Our challenge was the Barbara Perry family. Barbara, a single mother with six children, four to fourteen years old, was living in Bunceton in a rental home with boards crisscrossing the floor to walk on rather than fall through the rotting floorboards onto the ground below.

"The cycle of welfare had been played out with Barbara," Ken explained. "She was in need, and most importantly, the Lord had spoken to her, raising her up for the challenge of homeownership. The selection committee met, interviewed her, and took a leap of faith by unanimously recommending the sale of House No. 2 to Barbara Perry. She prayed and sought encouragement and guidance from her pastor and church family. Barbara was seized with apprehension about her need for the money to make her mortgage payments as well as feed and clothe her growing family. Barbara marched out to a mobile-home manufacturer in Boonville. Hiring was at a standstill and her application was destined to be filed for future reference. She gathered herself up and asked to speak to the plant manager. In a soft voice strengthened by her conviction to provide for her family, she informed him that she had been selected to purchase a Habitat house and she needed a job—now! She knew how to work and was an honest God-fearing woman who only needed a small opportunity to succeed in her effort to provide decent housing for her children. The plant manager was taken by her conviction," said Ken. "He found a job for Barbara. Today she is still employed there, is making her monthly payments, and has moved off her decade-long dependence on welfare."

Hand Up, Not Handout

Many Habitat homeowners have talked about how they do not feel demeaned in working to get a Habitat house. They like the partnership and the hand up instead of a handout. Their dignity is not assaulted in the process.

Pam Boleware is a single parent of four children in San Andreas, California. In the process of building her Habitat house, she became quite knowledgeable about construction. That enabled her to get a very good job with a local lumber company.

Gail Grow of Geneva, Ohio, offered the following observation: "My girls were sharing the single bedroom in our apartment. We were having way too many silly fights. And yet my apartment was what I could afford. I have never been good at asking for help. I felt that I should be able to provide for my daughters all by myself. My pride was messing up my daughters' lives. Habitat provided the answer. They accepted my gift of time spent in labor in lieu of a cash down payment. They call it sweat equity, but it earned me more than that. I gained valuable experience, confidence, and practical skills. Habitat allowed me to accept help without sacrificing my dignity."

Pride. Increased stability. Confidence. Practical skills. Experience and love. Affirmation. A solid roof and sturdy walls. Affordable payments. Privacy. All of these poured into the life of a family bring about transformation. And when families are transformed then the change flows out from each home into the neighborhoods in which they live too.

CHAPTER 4

Creating Neighborhoods in the United States

I would not have moved here when Habitat first came in.
But now look at it. It's safe and beautiful.
—NEW HABITAT HOMEOWNER

What happens when people love their homes? They naturally care about the block on which it sits. And when everyone on the block cares, then that is the kind of place where everyone wants to live. Imagine turning the corner onto a residential street of simple, decent homes. Each is neat and tidy, each shows signs of pride and joy. This ideal has been our motto from the very inception of the ministry of Habitat for Humanity—a decent house in a decent community for God's people in need.

We build good houses. That is basic. But we strive equally to build communities. We want to create, or re-create in many instances, strong and vital neighborhoods that strengthen people and help build solid families.

The importance of creating not only good houses but also strong neighborhoods has been recognized by the U.S. Congress. The 1949 Housing Act issued a call and stated the goal of a decent house in a suitable living environment for all American families. Achieving that goal has been elusive. As a nation, we seem to have done better at putting up housing than we have at creating "a suitable living environment." Consider the many unsuccessful government housing projects such as Pruitt-Igoe in St. Louis and Robert Taylor Homes in Chicago. The aim of the policymakers was simply to put up the largest possible number of units to alleviate the housing need of low-income families. Virtually no consideration was given to creating healthy and viable communities. The negative result was inevitable. Over time, artificial neighborhoods will always self-destruct, and these did.

We believe that every family on earth should have at least a simple, decent place to live. Implied in that bold goal is that the "simple, *decent* place" is in a "suitable living environment."

Of course, a good house in a suitable living environment is not a panacea for everything. Families must also be strong and supportive of adults and especially the children.

Heather MacDonald of the Manhattan Institute, writing in the *Housing Policy Debate Journal,* published by the Fannie Mae Foundation, stated, "Between a child and the neighborhood environment lies the family, and therein lies a crucial difference. While a lawless, chaotic neighborhood unquestionably poses great risks to a child's moral development, a strong family can act as a powerful buffer. Indeed, recent research suggests that family characteristics outweigh those of the neighborhood in predicting outcomes for youth."

This truth is why all local affiliates of Habitat for Humanity have family nurturing committees. We don't just build houses, place families in them, and walk away. Our intention is to help the families succeed as homeowners. We relate to the new Habitat homeowners in ways that strengthen them. We know that strong families are the best way to ensure stable and successful children. At the same time, we realize that if you can have strong families *and* have them living in a good neighborhood, that *doubles* the possibility of the families and especially the children doing well.

Building Houses Two Ways

Habitat for Humanity builds and renovates houses in two primary ways. First, many affiliates do *scattered site development.* This means that new houses are built or old houses renovated in various parts of a city, town, or county and then sold to low-income families. In both cases, whether building new homes or renovating old ones, the houses are scattered throughout the area. Typically, many of these houses are in low-income neighborhoods.

The second way that local affiliates build houses is *in clusters.* A plot of land will be donated to or purchased by the local Habitat affiliate and then developed by putting in streets, sidewalks, utilities, and so forth. Then a number of houses, anywhere from four or five to several hundred, will be built on the land.

In some instances, land or a number of derelict houses are acquired in an existing neighborhood. The streets, sidewalks, and utilities are already in, but the neighborhood has gone "downhill." Habitat comes in and builds or renovates a number of houses and thus turns the neighborhood around.

There are advantages and disadvantages to both methods—scattered site and cluster building. Many affiliates do both. The trend, though, throughout the United States and around the world is to do more cluster building. And the Habitat homeowners seem to prefer that way. They know, without having to worry, that their neighbors will love their home as much as they love theirs. Multiply that by the amount of houses on the block, and we have formed or re-formed a decent community.

The AREA study touched on this dynamic, reporting the following:

> One very positive factor currently mitigating homeowners' safety concerns appears to be Habitat's more recent pattern of developing either several houses along one street or entire streets of Habitat homes. In certain cases, where Habitat has developed a critical mass of homes in deteriorating neighborhoods, homeowners believe it has brought improvement beyond the boundaries of Habitat development. They credit Habitat with having reclaimed it as a safe and desirable place to live. Qualitative findings from interviews and focus groups suggest that Habitat's strategy of building in clusters and subdivisions has had a very positive impact on homeowners' neighborhood experiences.
>
> This is most evident in "difficult" neighborhoods—those characterized by general disinvestment and, frequently, by the incidence of illegal activity on adjoining streets. In these areas, homeowners cite an increased sense of security in being next to other Habitat homeowners. A number of respondents actually qualified their 'positive' responses about their neighborhood by specifying that they felt safer (in an area where they otherwise would have felt unsafe) because other Habitat homeowners were living on their street.

The study included some typical comments made by homeowners about their new neighborhoods:

> "Clustering makes the neighborhood a lot safer. We look out for each others' homes and kids."

"This area was always considered a rough area, with drug dealing and all, until Habitat homeowners moved in. . . . I love the cluster idea."

"This area's not so bad any more. We're [Habitat homeowners] putting pressure on the slumlords and drug dealers."

"I would not have moved here when Habitat first came in. But now look at it. It's safe and beautiful."

"Habitat offered me a home in that neighborhood and I didn't take it. But now I would because Habitat's made the neighborhood nice."

"Most people wouldn't live in this neighborhood, but I don't mind. It was really bad before Habitat."

"Habitat continues to build more houses in this neighborhood and is turning it around."

The Power of Being Chosen

There is a religious dimension to this matter of the homeowners' feeling of closeness or an affinity for each other, which results in a closer-knit community. Habitat homeowners are different from one another. In many instances, they did not know each other before they became Habitat homeowners. They work at many different kinds of jobs. Often they are of different races. What they all have in common is that they were *chosen* to have a Habitat house.

The families often put in their sweat-equity hours together as the houses are built. They frequently attend sessions on budgeting, home maintenance, and other such classes relative to homeownership. These common experiences set the stage for friendships to develop and for bonding to occur.

But why is all this a religious dimension? The concept of "chosen people" is deeply entrenched in Scripture. God *chose* the Hebrew people to be His people. The children of Israel are called *God's chosen people*. Jesus chose His disciples. They did not choose Jesus; He chose them. In the same way, the Habitat homeowners are *literally* a chosen people. That makes a difference and the evidence is there for all to see.

One place where a tremendous difference can be seen is the Winchester Park neighborhood in Memphis, Tennessee. Habitat for Humanity of Greater

Memphis has been building in that neighborhood since 1984. By early 1999, a total of eighty houses had been built or renovated. Celestine Hill and her family moved into their new Habitat house in 1992. "When I moved into my house, I was really excited about owning my own home for the first time," she said. "It makes you take more pride in the neighborhood in which you live."

Celestine constantly sought ways to improve her neighborhood. She and her neighbors are diligent in keeping empty lots free of trash and working together to reduce crime. "We still have crime," Celestine lamented, "but it has declined a lot, and we are still working on it. We talk to each other. We know each other. We share our telephone numbers so that if there is ever any trouble, we know how to get in touch."

Desperate to Get Out

Bruce Quinn and his family moved into their new Habitat house in Winchester Park in 1993. They had been living in an inadequate apartment in a bad neighborhood. Bruce and his wife, Wanda, were deeply worried about their three children. "We were desperate to get our children out of there," he exclaimed.

In Winchester Park, Bruce soon became a leader of the community. He was elected president of the homeowners association. "It is starting to look like a neighborhood again," Bruce said. He, Celestine Hill, and other neighbors continue to combat criminal activity and work steadfastly to improve the neighborhood. Meanwhile, his children have blossomed in their new and better environment. His eldest son, whom Bruce felt was on his way to being a hoodlum, is now in college. His oldest daughter is pursuing a practical nursing degree, and his youngest son is doing well in high school.

Joyce Collins, executive director of Habitat for Humanity of Greater Memphis, said that the Winchester Park neighborhood now has the lowest incidence of crime in the city of Memphis. Those Habitat families do make a difference!

Another neighborhood where Habitat families are making a difference is the Central neighborhood in Cleveland, Ohio. I was privileged to be in Cleveland on January 13, 1990, when ground was broken for the first house in that part of the city. At that time, Central was one of the worst crime-infested neighborhoods in the city. Now the situation is being reversed, according to Steve Kruger, family and neighborhood programs manager of Greater Cleveland Habitat for Humanity. Twenty-eight Habitat houses have been built there, and neighborhood residents played an active and vital role in planning and developing the new housing.

Turning Around the Neighborhood

Yet another neighborhood in Cleveland that is being impacted positively by Habitat is the Fairfax neighborhood. By early 1999, a total of thirty Habitat houses had been built, and the Fairfax neighborhood was showing dramatic signs of "turning around."

In the Fairfax neighborhood and other Cleveland neighborhoods, collaboration has been a key concept that has enabled Habitat's work to be enhanced. For example, in Fairfax, development of the neighborhood was done in concert with the Fairfax Renaissance Development Corporation, the Fairfax Project Council, neighborhood residents, private developers, and local institutions. The city of Cleveland, area foundations, and the local councilwoman were also vital components of the collaboration.

That fine team built fifty new houses, and more than a hundred more houses were repaired or renovated. This construction was the first in that neighborhood since World War II. The combined activity, led by Habitat for Humanity, was a big part of the bicentennial village observance of Cleveland.

A Medical Center for the Neighborhood

One quite wonderful aspect of the work was the realization of a long-held dream of Mount Olivet Baptist Church to build a medical center. Due to the dramatic transformation of the neighborhood, a hospital entered into a partnership with Mount Olivet Church and a medical center was built. The completed facility was dedicated in September 1997. Greater Cleveland Habitat for Humanity executive director Steve Frey said that they have ambitious plans to transform still more neighborhoods in the years to come.

First Habitat Homeowner Leads the Charge

Barbara Dunn, executive director of Paterson, New Jersey, Habitat for Humanity told the exciting story of the transformation of the Northside community led by the first Habitat owner in the community. "Plagued by decrepit housing, drugs, crime, and poverty" was how Barbara described the community when Habitat started working there in 1987.

Now there are more than eighty Habitat homeowners living within blocks of each other. "Habitat families," Barbara reported, "have taken leadership in reclaiming the neighborhood and making it a desirable place to live."

"The active involvement of homeowners began in the fall of 1993 when James Staton moved into his Habitat home right in the center of the Northside Community. He was elected president of a relatively inactive homeowners association," she said. "Within the first year he was publishing a monthly newsletter, holding regular community meetings, and taking action on issues of importance to residents of the area. The first priority was crime. James and the association set up telephone captains for each block of Habitat families and began working with the city on a community policing program. Community policing is now well established, and the homeowners association is a major partner in the effort. The Habitat families then worked long and hard to reopen a library that had been closed for three years. They were successful!" she said excitedly. "A bridge that had connected the neighborhood to downtown had been closed for more than nine years. Due to pressure from the homeowners, the bridge reopened in December 1997."

This is a wonderful image of a neighborhood's rebirth and re-acceptance by its community. Once cut off by a closed bridge, the dangerous neighborhood has reclaimed itself enough that the city leaders welcomed their access to downtown once more.

James Staton's leadership abilities continue to grow. He is continually strengthening the association, resulting in more and more families coming forward to join the effort. His copresident is Pedro Ruiz. "They are enormously hardworking individuals with a deep devotion to their families, church, and community. Pedro initially was quite reserved, but he has emerged as a strong team member in the work of the association," Barbara added. "For example, Pedro and James together make annual home visits to all the Habitat families in an effort to build support for the work of the association."

Barbara conceded that they have not yet "turned the corner" on making the Northside community a place where anyone would want to live and raise a family. "But," she said, "with God's help, the homeowners will succeed in taking back the neighborhood." By the year 2000, the one hundredth Habitat house will be built in Paterson and we pray that the one hundredth family will move into a community that is safe and beautiful."

Ten Condemned Shacks

Pensacola, Florida, Habitat for Humanity, one of the most dynamic affiliates in the world, has made a tremendous difference in several neighborhoods. But one transformation was especially significant. Kim Tully, development director of Pensacola Habitat, wrote about the changes:

> Over the course of a year and a half, Pensacola Habitat for Humanity dramatically transformed an entire city block. A row of ten condemned houses sat in the middle of the block. In addition to creating a dangerous eyesore, the shacks attracted crack dealers and vagrants and the police were frequently asked to respond to trouble there. Today, fifteen Habitat houses circle the block, including the milestone fifty thousandth house built worldwide by Habitat for Humanity.
>
> But more than the look of the neighborhood has changed because the lives of fifteen families living there have been profoundly affected as well. One homeowner is completing college and hopes to attend law school. Another homeowner had lost everything in a house fire. She and her four children moved into her brother's one-bedroom, one-bath house and the mother slept with two teenagers in the bedroom, two teenage boys slept on a mattress on the living room floor and the woman's brother slept on the couch. No one got much sleep—until she and her family moved into her Habitat house.

How Pensacola Habitat acquired the block is also an amazing story. The block of land was part of an estate that was in probate, and a family representative called and offered to sell half of the block for twenty-five thousand dollars. The Pensacola board of directors did not seriously consider the offer because most of the land on which they had already built nearly one hundred and fifty houses had been donated. They didn't like the idea of *buying* land. But the representative called back and reduced the price to fifteen thousand dollars. The day the matter was to go before the board, a Habitat supporter called and asked about any special needs. The president said that they needed fifteen thousand dollars for the property and six thousand dollars to demolish the ten condemned houses. The following day a check for twenty-one thousand dollars was received. At the end of the year, the same donor called and asked if they wanted the other half of the block. An affirmative reply brought a check for the other half of the block!

Habitat vs. Drug Neighborhood

My close friend Kenneth Henson Jr., an outstanding attorney in Columbus, Georgia, and volunteer executive director of the Columbus Habitat for Humanity affiliate for several years, described the exciting transformation of a neighborhood in that west Georgia city.

"One day in the spring of 1991, I was riding around the city with our building superintendent, Riley Middleton, looking for property for Habitat houses. I decided to make a turn down Thirtieth Avenue into a neighborhood which, in my childhood, had been known as Kendrick Quarters. I remembered as a child that Thirtieth Avenue was an unpaved street with shotgun houses. The area was known as Kendrick Quarters because the Kendrick family owned and rented most of the homes. Much to my surprise, most of the houses and apartments had been condemned and knocked down. Much of the area was vacant. The debris from many of the houses that had been torn down was piled up and overgrown with trees and weeds."

Ken contacted the Kendrick family through a mutual friend, and to his delight the family agreed to donate the property. There were enough lots for thirty-five houses.

The area was less than a hundred feet from an area where crack cocaine and other drugs were sold twenty-four hours a day. On Friday nights, Ken said, the intersection of Clarabelle Street and Thirty-first Avenue would have one hundred to two hundred people standing around drinking and using drugs. Crackheads and prostitutes walked up and down the street.

But that didn't stop Ken. He and his Habitat partners got busy. The first two houses were built in a blitz in August 1991. St. Luke United Methodist and St. Paul United Methodist sponsored one of the houses. A large number of Junior League volunteers and high school students came out to help. Ken told the families they were pioneers. He wrote to me about what happened next:

> After we finished those two houses, we were almost completely out of money. I remember when we first got started that you had told us to just start building. We decided to follow your advice and we started our third house in November. We did not have enough money to even dry in the house. But amazingly, the Junior League gave us some money and one day someone just walked up on the site and gave us a check for a thousand dollars. With these funds, we were able to dry in the house. The remaining funds appeared out of nowhere. We finished

this house in early 1992. After we completed this house, we were extremely blessed because churches and other organizations decided to help build houses. All of the downtown churches helped build and so did the homebuilders association, St. Francis Hospital, United Cities Gas, the *Columbus Ledger,* and many other companies and organizations.

Over the last few years, we shopped and bought more property in the area. We wanted to rebuild the entire neighborhood and get rid of the drug houses. We were able to get twenty-five more lots. We bought two drug houses and tore them down. In 1998 the district attorney's office confiscated a third drug house and we were ultimately able to buy it from the city, tear the house down, and then build a Habitat house where drugs had previously been sold.

We are almost finished building in this area. When we finish, we will have built over seventy homes in the Kendrick community area. We have learned from building the Kendrick community that Habitat works best when we go into a drug-infested bad neighborhood. We are able to buy the property at a reasonable cost or have it donated. We can then build a number of homes in one location. When we build these homes, we are not only rebuilding homes for families, but we are rebuilding and stabilizing the neighborhood.

A Playhouse

A touching story of a neighborhood of Habitat children comes from Mid-Yellowstone Valley Habitat for Humanity in Billings, Montana. Five Habitat homes were built in a row on South Twenty-ninth Street. The children from those families were close to the same age and played together. They even made a playhouse with leftover construction materials. And the "Habitat moms" are excited to be neighbors too. Two of the families are Native American, and two are white.

Reborn from a Hurricane

A story of courage, determination, and triumph comes from Kauai, the westernmost island of the state of Hawaii that was devastated by Hurricane Iniki on September 11, 1992. More than four thousand housing units were destroyed or seriously damaged in an area of only fifty thousand inhabitants. Even before the hurricane, though, there were serious housing problems.

Habitat for Humanity International took action. We sent Rick Hathaway to Kauai to start a Habitat affiliate. Rick is a dedicated staff person who has been with Habitat for Humanity since 1985, first as a volunteer in his local affiliate in Lynn, Massachusetts, and then as a volunteer in Americus. His biggest volunteer assignment was to lead the house-building portion of the 1988 House Raising Walk from Portland, Maine, to Atlanta. In 1989, Rick joined the Habitat International staff and became director of all U.S. affiliates. He served in that position until 1992.

In Kauai, Rick formed a group of dedicated people who were eager to begin the new Habitat affiliate. Within a year, the group was well organized, officially recognized by Habitat International, and had built fourteen houses.

Then Rick moved on to Korea to start Habitat in that country. LaFrance Kapaka-Arboleda became the executive director of the Kauai affiliate. In the years since, she has done a magnificent job of leading one of the most effective and productive affiliates in the country. By early 1999, Kauai Habitat had built a hundred houses, most on scattered sites. Even so, said LaFrance, there is a sense of greater community among the families. And in 1999, Kauai Habitat started building on a ninety-five-lot subdivision. I was privileged to be present in March to officially dedicate the land for the subdivision.

A Simple Announcement

Carolyn McCombs of St. Joseph, Missouri, Habitat for Humanity reported on a story of a neighborhood's changing simply by announcing plans to build a Habitat house. The Blessing family was living in a dilapidated trailer home on their own lot when they were chosen to have a Habitat house. When the announcement was made about their selection, another family decided to build a house on the adjacent lot to the west of the Blessings. After the Blessings' house was framed in, the Habitat office received a call from a man wanting to know if he qualified for a Habitat house. After some discussion, it was learned that he and his family lived in a derelict house on the east side of the Blessings' house. He was put in touch with another housing group that does rehabilitation work in the city. By the end of the summer in 1977, the Blessings' house was finished and the entire block was much improved.

No Nightclub for This Neighborhood

The Mid-Town neighborhood in Jackson, Mississippi, was filled with decaying houses, empty lots, trash, and other debris and was plagued by drug dealers and violence. Then Habitat for Humanity started building houses there in 1993. By late 1998, more than one hundred and forty-five Habitat houses had been built or renovated in the fifteen-block area near the state capital of Mississippi. The neighborhood was reborn. Among other things, the residents organized to keep their community in good condition. When a businessman wanted to build a nightclub in the neighborhood in 1997, dozens of homeowners appeared at the city's adjustments and appeals board to protest. Their efforts were rewarded when the application was denied.

Community Is an Enchanted Thing

Tonita Reifsteck of Decatur, Illinois, Area Habitat for Humanity reported that positive changes have occurred in every location where they built homes during the first ten years. She said that seven of their first sixteen houses were on adjacent lots. The neighbors in the surrounding houses, she observed, began to improve their houses. The whole neighborhood went through a vast improvement.

Susan Pyburn of Habitat of Ventura County (Oxnard, California) wrote movingly about an elderly African-American man who exchanged his plot of land for a three-bedroom house for himself, leaving room for two more families to live as his neighbors in modest Habitat houses. A small but vital community resulted. Here are Susan's eloquent words about the matter.

"Community is an enchanted thing for which no house plans can be drawn; the chief product of Habitat for Humanity, it is the stuff of friendship that makes neighbors out of people on the street, a fertile place for growing up or growing old, for teaching and lending and giving back. In Oakview, California, a tiny micro-community thrives," she reported. "On land too big for one and just about right for fourteen or so . . . three households stand together in the bond of friendship and a shared dream, representing the melting pot that is America's promise and the community that thrives where stable homes are born.

"People can and do get along, working side by side, neighbor to neighbor. If you want to see it work, come to Oakview and visit Mahoney Street. You might personally thank Charles Henderson, who turned seventy-six in 1998. He knows what he gave, and a deep smile glows in his eyes at the mention of his gift. He takes his place on the patriarchal porch, gazing out over the blooms now flourishing in the magical garden on the land he gave to Habitat."

The Opposite of NIMBY

Sometimes Habitat affiliates encounter what is called NIMBY (Not In My Backyard!). Neighbors are skeptical of low-income families, and they resist having Habitat houses built in their neighborhoods.

In Portland, Oregon, the Willamette West Habitat affiliate experienced the opposite of NIMBY. Here is the exciting story as recounted by Lem Taylor. "In the early 1990s our affiliate was struggling. We had finished our first two houses but were having a very hard time finding land on which to build. We were starting to feel pretty downhearted when Tony and Liz Jeffery stepped forward and answered our prayers. They had heard of our need for land and in essence told us, 'Here in my backyard.' As they put it, 'God gave it to us, so we are giving it back to God.'"

The Jefferys had bought a house with a big backyard at a price that in effect gave them the backyard as a bonus. After getting an easement from a neighbor, Habitat was able to build four houses "in their backyard." Later a neighbor's backyard was purchased, providing enough land for six more houses, giving the Jefferys ten families "in their backyard." That is quite a neighborhood.

Five-Hundred-Year Flood and More for Americus

In Americus, Georgia, home of Habitat for Humanity International, several neighborhoods have been revitalized, and others have been created. Since 1993, the excitement has been building and building and building. The Hope Community was launched that year with a spring blitz-build of twenty houses that were put up in a week. The milestone twenty thousandth house was in that build. Since then another twenty houses have been constructed on that land, which incidentally was donated by Americus's dedicated mayor, Russell Thomas Jr.

A little over a year later, in the summer of 1994, we had our celebrated Thirty–Thirty Thousand Blitz-Build. Thirty houses were built in just one week in June, and one was designated as the milestone thirty thousandth house built by Habitat for Humanity. That build was the largest ever done up to that time in the United States by Habitat for Humanity.

Within a month, two significant events transpired. While Jimmy Carter and two thousand volunteers built thirty houses in a week at the Cheyenne River Sioux Reservation in Eagle Butte, South Dakota, Americus endured a five-hundred-year flood. Raging waters came within a few feet of the new Habitat houses, but none were damaged.

During Holy Week in 1998, we launched a truly historic build of a new community in Americus. On the north side of town, in a beautifully wooded area, eleven hundred volunteers erected the first twenty houses in a special build called the Easter Blitz-Build.

At the end of the week, on Good Friday evening, a glorious service was held at First United Methodist Church with celebrated speaker Tony Campolo as the featured preacher. He thrilled the hundreds of assembled volunteers with his famous "It's Friday, but Sunday's Coming" sermon.

Then, at daybreak on Sunday, we held a wonderful sunrise service at the construction site, and it was attended by a huge crowd, including the home-owner families and scores of the volunteers who had stayed over to be a part of that service. The excitement and joy of the people were truly electric. All who attended sensed the presence of God among us. It was like a foretaste of the kingdom because there was so much love in that crowd.

Linda gave a thoughtful devotional message. Beautiful voices sang and several others spoke, including a young pastor whose family was moving into one of the houses. He told the hushed crowd that if drugs were allowed into the community, it would be known as a drug community. If alcohol was allowed, it would become known as an alcohol community. But, he exhorted, if God was the head of the community, it would be known as a godly community. "Let God be the head of our new community!" he shouted. The loud amens resounded throughout the receptive audience.

At the conclusion of the service, everyone marched to the entrance of the new neighborhood to unveil a plaque that revealed a burst of sunshine below which was inscribed, Easter Morning Community. And beneath that boldly written name was the slogan, "Every house a sermon of God's love."

This unique new subdivision—which will have a total of one hundred and forty-two houses, all of which will be finished and occupied by the year 2000—is part of Habitat's overall campaign to get every family in our area into a good house by Building on Faith week in September 2000.

During Holy Week of 1999, another twenty-five houses were built by more than fifteen hundred volunteers. A third Holy Week build is planned for 2000. The remainder of the houses are being built on a more normal schedule.

Big, Big Neighborhoods

How big can our Habitat neighborhoods be? Two of the largest communities in the United States being created by Habitat for Humanity are the one hundred ninety-six house South Ranch subdivision being developed by Valley of the Sun Habitat for Humanity in Phoenix, Arizona, and the Jordan Commons one hundred eighty-seven house development in Homestead, Florida. The Florida subdivision was started by Homestead Habitat for Humanity, which was absorbed by Greater Miami Habitat in 1997. It is named after Clarence Jordan, my spiritual mentor and the man who, with his wife, Florence, and another couple founded Koinonia Farm, the birthplace of Habitat for Humanity. Both of those new communities are full of innovative features designed to enhance the lives of the residents. They are exciting places to live and wonderful places to visit.

Atlanta Habitat for Humanity has developed or revitalized numerous neighborhoods, including ones with descriptive names such as Cabbagetown, Mechanicsville, and Reynoldstown. One community on the west side of the city was targeted by a Jimmy Carter blitz-build in 1988. Twenty houses went up in a week, and that event set in motion a process which greatly transformed the whole area. Other developers came and built a number of other houses so that a neighborhood that was once filled with derelict houses was changed into an attractive part of the city. Jack Kemp visited that neighborhood with me and other housing leaders on his first day in office as secretary of housing and urban development (HUD) under President George Bush.

Atlanta Habitat for Humanity continues to be a tremendous agent of change. It has completed or renovated five hundred houses and continues to build sixty-five houses a year.

Just northwest of Atlanta, headquartered in Marietta, is Cobb County Habitat for Humanity. That affiliate, started by our dear friend Chrys Street, had built one hundred twenty-two houses by the end of 1998, mostly in three neighborhoods. The biggest neighborhood is called Whispering Glen in the town of Powder Springs. There are eighty lots in the development. As of early 1999, fifty of the houses have been completed. The remaining thirty will be built over the next couple of years. Beggs Court, named after our good friends and dedicated Habitat partners Linda and Philip Beggs, is in Marietta and has fifteen lots. Two houses had been completed there by the end of 1998, and eight more are scheduled to be built in 1999.

Also in Marietta, in the very center of the city, Cobb County Habitat for Humanity collaborated with two other organizations to build and renovate homes in an area near Roosevelt Circle. Habitat built and renovated twelve houses and demolished a drug house. That neighborhood had the highest crime rate in the city. After the building and rebuilding work, the crime rate went down significantly.

Land for twenty-nine more houses has been purchased in the town of Mapleton. Those houses will be built within the near future, creating yet another Habitat neighborhood in Cobb County. Cobb County Habitat, incidentally, is building twenty-five houses a year locally and many more in Uganda through its faithful tithing. We estimate that Cobb County Habitat has built one hundred houses in Africa through its tithe contributions.

Charlotte Habitat for Humanity is another dynamic affiliate that has mastered the art of transforming neighborhoods. In 1987, the annual Jimmy Carter work project was held in Charlotte. Fourteen houses were built in Optimist Park in a week. Charlotte Habitat had already been working there for a year or so, but much remained to be done. In the years since, scores of additional houses have been built and the community has been completely changed. A total of sixty-three Habitat houses now stand proudly in that lovely neighborhood. In 1988, Charlotte Habitat went across the creek from Optimist Park into the run-down neighborhood of Belmont. In nine years, one hundred and seventy more houses were built or renovated there, resulting in another transformed neighborhood.

In fifteen years, Charlotte Habitat has built a total of three hundred and sixty houses, and they continue to build scores more every year. More and more neighborhoods are being changed in the process. And as they faithfully build in Charlotte, through generous tithing, an estimated four hundred houses have been built in developing countries.

That's one of the wonderful ways that neighborhoods help neighborhoods. We have a system within Habitat in which one affiliate "tithes" to help affiliates in other countries build Habitat neighborhoods. Its worked amazing feats in people's lives and neighborhoods across the country and world.

Jacksonville, Florida, Habitat for Humanity, locally known as HABI-JAX, is truly an incredible operation. With a dedicated and effective board and with a talented staff led by Frank Barker, that affiliate keeps going from victory to victory. In 1998, they built a hundred houses. In 1999, their goal is one hundred and forty houses, and in the year 2000 they plan to build two hundred houses! A hundred of those houses in 2000 are to be built during Building on Faith week, when all Habitat affiliates plan to build a total of ten thousand houses around the world in connection with the dedication of our one-hundred-thousandth house that week.

I was in Jacksonville in June 1998 to help HABIJAX celebrate their tenth anniversary. The previous week they had blitz-built twenty-one houses. It was such a joy to ride around the neighborhood where they were building and see what a change was being made. When this book went to press in 1999, Jacksonville Habitat had completed a total of three hundred and forty-eight houses, and their generous tithing has built many more houses in developing countries.

Volunteers labor under the blazing Houston sun at the 1998 Jimmy Carter work project in Texas.
Photo by Robert Baker.

Signing Their Happiness

Scores upon scores of other affiliates all across the United States have built or rebuilt neighborhoods. A true *master* of neighborhood development and transformation is Twin Cities Habitat for Humanity in Minneapolis and St. Paul, Minnesota. Under the dynamic leadership of Stephen Seidel, Twin Cities Habitat has created a series of chapters that work throughout the two cities. They have built a total of two hundred and fifty houses as of early 1999 and continue to build forty-five to fifty a year. By 2003 they plan to be building eighty houses a year. In 1995, Twin Cities Habitat was chosen to build the milestone forty-thousandth house, and I was present for the happy occasion of dedicating that house. The house was built for a deaf couple, Theodore and Louise Williams, and their five children and had some special features, such as a flashing light as a doorbell. With sign language, Louise expressed their great joy at being able to move into a good home on terms they could afford to pay. Twin Cities Habitat has also been one of the most faithful and generous tithing affiliates in the country.

Pioneers Still Building and Building

Immokalee Habitat, now known as Habitat for Humanity of Collier County, was the first Habitat affiliate in Florida and the third in the nation. I was privileged to be present for the birth of that pioneering group with Bob and Myrna Gemmer and Bob and Amy Olsen. Bob and Amy were volunteer workers with the Church of the Brethren Volunteer Service program. Bob was a recently retired professor of engineering at Penn State University. The Gemmers and Olsens had invited me to Immokalee, and I went with Linda in January 1978. We held the historic meeting that is considered the founding session of that remarkable organization on January 29. The meeting was held in a small concrete-block house, typical of those occupied by migrant farm workers in that deep south Florida town.

More than a hundred people came to that meeting. Not everyone could get into the house, so they peered through windows and doors. I showed some slides of our work in Africa and at Koinonia in south Georgia. Then I talked about the concepts of the work, with houses built largely by volunteer labor, including the prospective homeowners. I told about the biblical principle of no interest and of our challenge to those with excess to share with those who had too little. I concluded by saying that the same building we were doing in Africa and south Georgia could be done in Immokalee to help the hundreds of families there who desperately needed a good and decent place to live.

When I finished my talk, a little girl, about four years old, who was sitting on the opposite side of the room, started crawling over people to come to me. When she made it to my chair, without saying a word, she crawled into my lap and hugged me. It was a magical moment. All of us present that night saw that gesture as a sign from God that we were on the right track.

It took a year following that first meeting to get the first house built, but once begun, the building never slowed down. In the twenty-plus years since that first meeting, more than two hundred and sixty houses have been built and thirty to thirty-five more are being put up each year. And even though the need is so great in Immokalee and elsewhere in Collier County, the affiliate has faithfully tithed to build houses in other countries.

Kansas City, Missouri, is another pioneering affiliate in the United States. John and Mary Pritchard were the founders of that group, the first

Habitat affiliate in Missouri and the sixth in the nation. I was privileged to be present in October 1979 to break ground for their first house in the Mount Hope neighborhood. In the twenty years since, they have built fifty-six houses there and another eighty-one in the adjacent Boston Heights neighborhood. As the new houses have gone up, the happy residents have seen the overall quality of life also rise and crime has steadily declined. Kansas City Habitat also faithfully tithes, and that money over the years has built an estimated two hundred and ninety houses in developing countries.

Still another pioneering affiliate that has done such a great job of building good neighborhoods is Sea Island Habitat for Humanity, based on historic John's Island. The first Habitat affiliate in South Carolina and the fourth in the nation, that dynamic group was led for eighteen years by Jim Ranck, a quiet Mennonite pastor with a fervent heart of love and concern for the poor. Jim and his dedicated associates have built sixty-six houses in their twenty years as a Habitat affiliate. Their two big neighborhoods are Habitat Place and Taylor Place, with fifty-three houses, and a newer subdivision with thirteen houses. They continue to build fourteen houses a year. Their tithing has also been excellent and has made possible the building of forty-five more houses in developing countries.

In Texas numerous affiliates have done a fantastic job of building some great neighborhoods. San Antonio Habitat for Humanity, the first Habitat affiliate in the United States, has grown steadily over the years and has developed and transformed some really great neighborhoods, mostly on the west side of the city. In November 1996, Linda and I were privileged to be a part of their twentieth anniversary celebration. In a two-week blitz, involving forty-six hundred people, twenty houses were built in a beautiful new neighborhood.

On Sunday, while there, I preached in Trinity Baptist Church, a sponsor of one of the twenty houses. A wonderful little irony happened there. The name of the mother of the family that would live in the sponsored house was Anna Bautista, a Catholic. Bautista means "Baptist" in Spanish. That is an excellent example of the theology of the hammer in operation. San Antonio Habitat continues to bring people and churches together, and it continues to transform lives and neighborhoods in more and more neighborhoods in San Antonio and, through tithing, in other countries.

Deep in the Heart

Dallas is yet another city where Habitat has built and transformed neighborhoods. In 1997, Dallas Habitat was chosen to build the milestone sixty-thousandth house. It was a part of a twenty-one-house blitz that brought hundreds of churches together during Building on Faith week. I was privileged to be present for that historic occasion. With the mayor and other dignitaries, we erected the first wall of that special house as other volunteers cheered. They were cheering not only for that special house, but for the change that was coming to that neighborhood and the city. Dallas Habitat has built one hundred sixty-four houses in Dallas as of the end of 1998, and even more in developing countries through tithing.

Four Neighborhoods in a Week

Houston Habitat for Humanity was the site of the historic one-hundredth house built in June 1998. Six thousand volunteers joined former President Jimmy Carter and his wife, Rosalynn, to build four neighborhoods in a week. No organization in the United States had ever done anything like that before. The weather was unbelievably hot, with the *cool* days being one hundred degrees. Even so, the houses got built, and during the following week, the homeowners moved into their brand-new homes.

Such a huge build was an enormous challenge to the leadership of Houston Habitat, but executive director Michael Shirl, board president Harvey Clemens, International board member Carl Umland, fund-raising chairman James Calaway, and other dedicated Houston Habitat leaders made it happen. And they have done excellent follow-up with the families to ensure their success as homeowners and to promote the establishment of strong and viable new neighborhoods. Beyond the huge blitz, Houston Habitat continues to build and renovate more houses each year as they support and encourage the work in developing countries through faithful tithes.

A Million-Dollar Tornado

In my native state of Alabama, Habitat affiliates are now transforming neighborhoods in more than thirty towns and cities, and state director Lou Hyman is determined to see the number of affiliates keep climbing in the years ahead until every town and county in the state has a Habitat affiliate. Montgomery Habitat has done a good job of building neighborhoods. In 1995, I went to Montgomery during Building on Faith week to break ground for a new subdivision for sixteen houses. In 1998, I was back to help dedicate the last house. It was inspiring to see the transformation that had been brought about in that new community.

Birmingham Habitat for Humanity was building in several neighborhoods when, during Holy Week 1998, a series of deadly tornadoes devastated several poor neighborhoods in west Jefferson County. An estimated 459 houses were destroyed and another 643 severely damaged. Thirty-six people were killed. Even though Habitat for Humanity is not a relief agency, Birmingham Habitat felt compelled to act. So many families lost everything and were too poor to rebuild their lives without outside help.

Jan Bell, executive director of Birmingham Habitat, called me at the work site of the Easter build in Americus. "Millard," she said, "we've got to do something to assist in the devastated areas. Can Habitat International help?"

Celestine Hill cuts the ribbon at her new home dedication service in
Winchester Park in Memphis, Tennessee, with sponsor partners from
Telephone Pioneers of America, Grace St. Luke Episcopal Church, and
Church of the Holy Communion.

Fortunately, several of our staff were there. Sandi Byrd, our senior vice president of development and communications, was on site along with David Williams, our senior vice president of administration, and Ted Swisher, director of all U.S. affiliates. Roger Craver and other top leaders of our direct-mail consulting firm were there too, helping to build the twenty houses at the blitz site. I convened an emergency meeting in the front yard of one of the houses, and a decision was quickly made: We've got to help. A mailing was planned and a commitment made to Birmingham Habitat to raise a million dollars.

The effect was electric. Within a few weeks we had in hand more than $800,000 and Birmingham Habitat had gotten commitments for another million dollars from Birmingham sources. Their biggest partner was Alabama Power Company, whose president and CEO, Elmer B. Harris, contracted a good case of Habititus. The company helped sponsor and quickly built the first two houses for tornado victim families. Additional help was forthcoming from Alabama Power and other companies, organizations, and churches. By September, four houses had been completed and two more were under construction. Twelve additional homes were finished within a year of the tornado disaster.

On September 25, I went to Birmingham for a day of activities organized by Jan Bell and her dedicated staff in conjunction with Alabama Power Company. The key event was a breakfast sponsored by Alabama Power and attended by representatives of the business community and religious leaders of the city. Jack Kemp was the featured speaker. After his speech, I issued the challenge to follow the lead of Americus and Sumter County to set a date to eliminate *all* poverty housing in Birmingham, using the rebuilding of the tornado-devastated neighborhoods as the catalyst to launch the effort. I am optimistic that this budding initiative will continue to strengthen in the months and years to come. Jan Bell's goal is to be building a hundred houses a year in Birmingham within the near future.

Anniston Initiative—All Substandard Houses Gone

Anniston, a city of twenty-nine thousand people roughly halfway between Birmingham and Atlanta, has already officially launched their campaign to eliminate all substandard housing. A delegation from Anniston Habitat, led by executive director Bill Wright, came to Americus during the Easter build. They were inspired by what they saw.

Clive Rainey, senior officer in our development department, had been meeting with the Anniston affiliate, Habitat for Humanity of Calhoun County, for several months. Clive has taken on the challenge of launching a whole series of towns and cities to eliminate poverty housing, following the Americus–Sumter County model. The new initiative is being called the Twenty-first Century Challenge. By mid-1999, more than fifty towns and cities had enrolled in this innovative campaign.

Following the visit to Americus, the Anniston folks, in full cooperation with the mayor of Anniston, the city council, the chamber of commerce, and other business, political, and church leaders, organized a big kickoff rally to announce the Anniston Initiative to eliminate all substandard housing in their area. The meeting was planned for Friday evening, August 14, 1998. Clive and I, along with Americus mayor Russell Thomas and his wife, Andrea, attended. The excitement was incredibly high. The local paper had been running articles and a strong affirming editorial. That evening at the Anniston City Meeting Center, nearly three hundred people were in attendance, including the mayor, the entire city council, the president of the homebuilders association, most of the CEOs of local companies, and numerous pastors and other church leaders.

Everyone was positive about the historic step of organizing to get every family in the city and county into a good and adequate house. Bill Wright informed the crowd that a date would be set for the accomplishment of the goal by the end of the year.

The following morning, we joined a hundred volunteers to frame the first house of the new initiative. The house sponsor, a company named Solutia, was well represented on the site by Blake Hamilton and a score of other company employees. I told the people in my devotional that morning that Solutia was the perfect sponsor for that first house because their name sounded like Latin for "solution." They had the solution, and if others would simply follow their example, the goal would be met. In December, as promised, Bill Wright announced that by 2015 all poverty housing would be eliminated in Anniston. He estimated that five hundred new houses would be needed to do the job.

Yet another city in Alabama, known throughout the nation for its racial problems, is Selma. Habitat for Humanity is working there with great effectiveness, especially in developing neighborhoods and bringing the races together to accomplish something good and positive for the city. During Building on Faith week in September 1999, a forty-six-house subdivision will be launched with the building of the first twenty houses. Longtime Habitat volunteer Alan Riggs of Kingsport, Tennessee, is directing the blitz-build.

In the Wake of Rodney King

Another city that has become divided along racial lines is Los Angeles. People the world over came to know the name of Rodney King when he was beaten by police officers in 1991. The beating was captured on video and broadcast repeatedly for months. When the officers who beat him were acquitted, black neighborhoods erupted in violence. The scars from that violence are still healing in that city.

In the wake of that violence, President Carter in 1995 led more than a thousand volunteers into the community of Watts to blitz-build twenty-one Habitat houses. A beautiful island of tidy and solid houses was left by those dedicated builders. Los Angeles Habitat continues to build on the legacy of the Jimmy Carter Work Project. Executive director Barry Witmer and other staff and board members are doing a fine job of changing mistrust to trust, hopelessness to hope, and despair to rejoicing as they keep on hammering out faith and love in the City of Angels.

A Habitat State

An entire state that is building good Habitat neighborhoods is Michigan. The most dynamic state leader of Habitat for Humanity in the nation is Ken Bensen, a tireless United Methodist pastor in the state capital of Lansing. For years, he has crisscrossed the state, planting new Habitat affiliates. The result of his efforts is eighty-plus Habitat affiliates in Michigan, more than any other state.

Ken is constantly organizing special projects to boost the work. For example, in the summer of 1997, he organized a statewide blitz-build that put up one hundred and seventeen houses during a two-week effort. June 1999 saw a one hundred and eighty house two-week blitz-build. It was being called the Jack Kemp Blitz-Build. You can be assured that these builds and the more routine building being done all over Michigan are transforming neighborhoods. By the end of 1999, Ken expects to have a Habitat affiliate for every *inch* of the state of Michigan.

Another outstanding state leader is Mindy Shannon-Phelps of Kentucky. She coordinated the Kentucky portion of the 1997 Jimmy Carter work project and the Echo Build that involved scores of other housing groups working in partnership with Habitat for Humanity. Mindy, like Ken Bensen, is always promoting events to boost the work of Habitat across the state. For example, she plans to have scores of affiliates in Kentucky building at midnight to usher in the new millennium, joining with many other affiliates in the United States and around the world that are doing the same thing.

Habitat Magic and the Chicken Man

Just as Ken and Mindy are doing a fabulous job in Michigan and Kentucky, Terri Bate is practicing a special form of "Habitat magic" in Canton, Ohio. Terri is the incredibly talented and committed executive director of Canton Habitat for Humanity. Under her leadership, that affiliate has dramatically transformed several neighborhoods. She has secured the support of scores of churches and the backing of dozens of corporations and other organizations in Canton. Because of her outstanding work, Canton Habitat for Humanity was chosen to build and dedicate the milestone seventy thousandth house. I was privileged to be present for the seventy thousandth house event during Building on Faith week in September 1998. That milestone house was also Canton Habitat's one hundredth house.

A very humorous thing happened at the build site for the seventy thousandth house. Ohio Governor George Voinovich was there helping with the construction work. He was nearing the end of his second term as governor. Not being able to run again for that office, he was an active candidate for the U.S. Senate. His opponent was a woman from Cleveland. She had challenged the governor to debate her, but he had refused. As a way to embarrass Voinovich into debating her, the woman hired a man to dress up in a chicken costume and heckle the governor whenever he made public appearances.

The "chicken man" showed up at the Habitat site and started heckling Voinovich. Canton Habitat for Humanity officials confronted the man, claiming that Habitat for Humanity was not a political organization and asking him to take off his costume, put down his sign, and help Voinovich with the work. To the amazement and delight of everyone, he did just that and worked all afternoon helping to build the house.

As good as Terri is in Canton, her good work is not restricted solely to that local area. She constantly assists and encourages sister affiliates in Ohio and beyond. Terri serves on the women's build steering council, which encourages women to be involved in building Habitat houses in the United States and around the world. She is a faithful and tireless promoter of the overseas work. The faithful tithing of Canton Habitat has built two hundred and thirty-one houses abroad. Terri has organized three Global Village work teams that have worked in Nicaragua, building more Habitat houses and transforming additional neighborhoods. One of her teams was in Nicaragua when Hurricane Mitch hit the country. They were marooned in a village for a week before getting out. Sometimes the sacrifices of volunteers are substantial, but the rewards are so incredible for the people who now have houses in that country because these workers risk their lives to help. That neighborhood in Nicaragua will never forget their Habitat neighbors from the north. And that connection is an important one. Because as Habitat for Humanity is building good houses and strong neighborhoods throughout the United States, it is also building good neighborhoods all around the globe.

CHAPTER 5

Building Neighborhoods Around the World

*One by one several people stood to speak, but much to our
surprise not one mentioned anything about their new houses.
Instead, each spoke of their new community.*
—ANDY KRAMER, INTERNATIONAL HABITAT PARTNER

A concrete block. Another and another and another. A doorway.
Windows. Shelter. A simple, decent house in most of the neighborhoods of the world is not much more than that. The costs are so low, the need is so great, and the pride is so special that whole neighborhoods are being built everywhere. House by house, country by country, Habitat is creating good neighborhoods worldwide.

With the United Nations as our ally, Habitat for Humanity boldly states that *every family worldwide* should have at least a simple, decent place to live and in a *good neighborhood*. The most amazing thing about this statement is how little such a house costs in terms most Americans understand. How much does such a house cost? A Habitat house in a developing country costs an average of two thousand to three thousand dollars. Compare that to the average cost of a Habitat house in the United States, which is now forty thousand dollars and up.

Habitat houses are being built for immigrants from Mexico, El Salvador, Guatemala, Haiti, Vietnam, and other such countries, and I am thrilled that our local Habitat affiliates are building houses for these immigrant families. The Bible is clear that we should show love and kindness to the foreigner in our midst. But there are good reasons we should help people in other countries and attempt to give them help to stay where most of them would rather stay—in their homelands.

Why Build Abroad?

From time to time, I am challenged by people who question why we build around the world. I always respond in two ways. First, and most importantly, God's love knows no boundaries. God is not an American citizen. Jesus never held a U.S. passport. God's Word is clear. Our love should have no bounds. Jesus taught that we should love even our enemies. Surely, if that is the case, we should love people who live in Mexico. But what about Cuba? Should the people of that nation be shunned?

As an unabashedly and openly Christian ministry, we should always espouse the love ethic of the Carpenter of Nazareth and build for needy people regardless of race or religion or what country they might live in. Likewise, we should welcome people to help in this ministry from all churches and no churches. All are invited to be a part of the solution, and all families in need are eligible for assistance. This borderless philosophy extends across religious and national lines. We build at home and around the world.

The second reason for building worldwide is a more practical one. We live in an age of instantaneous communication. A faxed message or an e-mail can reach the other side of the world almost as soon as it is sent.

People all over the world *know* that America is the land of opportunity. They know about our standard of living, which is far above most nations of the world. The United States has deficiencies, to be sure, and substandard housing is one of those glaring deficiencies, even with Habitat for Humanity and other groups and organizations, including governments at various levels, working steadily to solve the housing problem. The rest of the world is aware of what is going on in this country. News from the United States is broadcast hourly by satellite news media all around the world. The American dream is known and envied around the world, because we are perceived as wealthy in comparison—all of us. As someone said, our poor are considered rich by most of the world's impoverished people. Much of the problem is the displacement of these people from their villages and huts due to economic factors, so more and more people want to migrate to the United States every year. There are no great hordes of people clamoring to relocate to Mexico, no long lines of U.S. citizens at the doors of the Liberian Embassy desiring to become citizens of that country.

There are, though, hundreds of thousands of people in countries around the world who want to move to the United States.

How do we deal with that problem? The poem at the base of the Statue of Liberty boldly proclaims, "Give me your tired, your poor . . ." but as big as our hearts, as Americans, continue to be, there are now limits on how many immigrants this or any country can absorb. One U.S. politician in the last election proposed a high wall to keep out foreigners. Is that the answer? Is that a Christlike solution? Hardly.

A better way, it seems to me, is to help elevate the living standard of people *where they live.* Most people in the world would prefer to stay where they are. They prefer to be with family and friends in a culture with which they are familiar. They only pick up and move when life becomes so intolerable that they are willing to risk the uncertainties of the journey and the unknowns of a foreign land. So *enlightened self-interest* plus the powerful motivation of the love ethic of Christ should propel us to be more bold in reaching out to help the foreigner where he or she lives.

Two Thousand Goes a Long Way

But there's also another good reason, and it has to do with what two thousand dollars can buy in the rest of the developing world. The economics of the situation clearly dictate that we could help *more* people *where they are* because the cost is so much less to do so.

At this moment, Habitat is building in approximately one thousand locations in more than fifty developing nations, in more than fifteen hundred towns and cities in the United States, and in a half-dozen developed countries. And new affiliates and more countries are being added every year. Our intention is to launch this hammering ministry in every nation on earth and introduce the idea in each of those countries that every family should have a good and decent place to live in a suitable living environment.

We fund this work around the world in a variety of ways. First of all, Habitat for Humanity International raises money for the worldwide work and makes regular transfers of funds to the various countries. This money comes from individuals, churches, companies, and other organizations, primarily in the United States, but increasing money is raised in other countries too. For example, we now have a strong national Habitat for Humanity organization in the Netherlands. Since there is virtually no poverty housing in the Netherlands, the Habitat people there recruit volunteers and raise money for work in other nations.

There is a budding interest in forming a Habitat for Humanity national organization in Switzerland to raise money and recruit volunteers for other countries. We expect to form many other such national Habitat organizations in developed countries around the world in the years to come to help fund houses in developing countries.

Funding the World Dream

The local Habitat affiliates in *all* countries raise money from local and national sources to fund the building of the Habitat houses in their respective countries. In fact, Habitat International usually ties our transfers of funds to how much has been raised locally. For example, in some countries we transfer a dollar for every half dollar raised locally. In other, poorer countries the ratio can be as high as nine U.S. dollars for every dollar equivalent raised locally. Government funds in some countries are also available, particularly for land acquisition and infrastructure development.

In addition to the above-mentioned ways of acquiring funds for the international work, tithing is a worldwide method of generating money. Every Habitat affiliate *throughout the world* is *expected* to give 10 percent of locally raised funds, except for money specifically designated *solely* for work within that country, to work in another nation. In 1998, that sum of money sent in by the local affiliates for work abroad was in excess of six million dollars. The amount for 1999 is expected to be at least one million dollars higher than the 1998 tithe gifts.

The tithe contributions have gone up consistently every year since the inception of this ministry. At first, only the United States, Canada, and other developed countries participated in the tithe program. In 1993, though, the so-called Entebbe Initiative expanded the tithe expectation to all countries where we have work. Even the poorest country is now expected to tithe to another country. The response from the various countries has been phenomenal. They greatly appreciate the opportunity to give as well as to receive. As the tithe program continues to grow, more and more houses will be able to be built and more neighborhoods developed and renewed in nations around the world.

The First International Story

The very first work done in this house-building ministry outside the United States was in Africa. Linda and I moved to central Africa to the nation of

Zaire (now the Democratic Republic of the Congo) with our four young children, Chris, Kim, Faith, and Georgia, in 1973 to work as missionaries with the Christian Church (Disciples of Christ). We were assigned to do development work with the Disciples of Christ community of Eglise du Christ au Zaire (The Church of Christ of Zaire) in the Equator Region. Our home and base of operation was Mbandaka, the capital city of the region.

Upon arriving in Mbandaka we found a rundown "block-and-sand" operation that had been purchased by the church from a Belgian business-man. The enterprise consisted of a tugboat and a couple of barges bound together by a platform on which was mounted a crane. On the shore was a conveyor belt used for off-loading the sand from the platform. Nearby was a small brick office and an adjacent open shed, under which were two small block-making machines, a cement mixer, and some wooden pallets for use in drying the concrete blocks after they had been compacted by the machines. And there was an old dump truck that had been used to haul the sand and blocks.

The business had thrived in earlier days by digging sand from the river bottom, putting it on the barge platform, hauling it to shore, and off-loading to the dump truck. Some of the sand was sold to customers in town, and the remainder was mixed with cement to make and sell concrete blocks.

But when we arrived, nothing was being sold. Nearly everything was bro-ken down. The conveyor belt had been patched so many times you couldn't tell the conveyor belt from the patches. The truck barely ran. One of the block-making machines was broken and the other was barely functioning. The pallets were falling apart.

I was excited, though, to have even this broken-down enterprise. I could see that if we could get it to run again, funds could be generated and most importantly, blocks could be made and used to build houses.

The full story is told in my first book, *Bokotola,* but let me simply say that we rejuvenated the block-and-sand enterprise. Hundreds of tons of sand were brought in to shore. Some was sold to the public, but most of it was used to make concrete blocks. Some of those blocks were sold, but, again, most were used to build houses.

With the leader of the church community, I contacted the regional com-missioner (the equivalent of a state governor in the United States) and con-vinced him to give us the old segregation strip in the middle of the city that had separated the Africans from the Europeans when the country was a Belgian colony. The Belgians had called this wide strip of land "the health strip." But the Africans, seeing the land from their perspective of poor housing, dirt roads, and a neglected part of the city, called it "Bokotola," which meant in the Lingala language, "A man who does not like other people."

We got to work on this land and laid out streets for one hundred and fourteen houses. By the time we left Africa in 1976, all one hundred and fourteen houses were under construction and a score of them were complete and occupied.

In the meantime, Linda and I visited the small village of Ntondo, which is on the eastern shore of Lake Tumba, some ninety miles south of Mbandaka. There we helped launch another house-building program to put up three hundred more houses. On the eve of our departure from Zaire in mid-1976, a huge celebration was held at the central park and playground at the building site in Mbandaka. Top church and government officials were present. The crowning part of the program was the official renaming of the property. Church officials proclaimed that the name Bokotola was no longer appropriate. The land had been transformed. Hence the old name was being rejected. The land and the emerging neighborhood would henceforth be known as Losanganya, which means "a place of reconciliation, where people love and respect one another."

Losanganya—Place of Reconciliation

Following the service, Linda and I returned to our house. We were eager to continue packing because our departure for the United States was the next day. To our surprise, though, the bishop's car pulled into our driveway. "You are needed at the church," the driver said. We were puzzled because we thought the program was over, but we loaded into the Volkswagen and rode to the church.

The sanctuary was packed with people. Linda and I were summoned to the front, and a service commenced in which we were given official African names. "From this day forward," the bishop said, "you will be known by us as 'Mr. and Mrs. Losanganya.'" It was a powerful and unforgettable moment for us. In the ensuing years, all one hundred and fourteen houses were completed at Losanganya and so were the three hundred houses at Ntondo. More streets were added and still more houses built at Losanganya so that eventually nearly two hundred houses were built there. Further expansion occurred in southern Equator Region by building in Bikoro, a town a few miles from Ntondo, and all along the road in villages from Mbandaka to Bikoro.

In 1995, Linda and I returned to Equator Region and were thrilled to see the community at Losanganya thriving. We also saw the hundreds of houses along the road to the southern part of the region, and then visited many of the hundreds of houses in both Ntondo and Bikoro. Other expansion has occurred over the years, including a new community of more than one hundred families in an area of the national capital of Kinshasa called Mount Ngafula. In all, by early 1999, more than two thousand Habitat houses had been built in that country, and the work continues in spite of almost unbelievable economic and political problems.

The various neighborhoods that have been built are improving despite the continuing problems in the country. The Mount Ngafula community, for example, has also constructed a church and a school. And the people have also organized a fish-farming community enterprise at the foot of the hillside on which the houses are located.

Fifteen Countries in Africa—Scores of Tidy Houses

From the humble start in Zaire, Habitat for Humanity has grown to fifteen countries in Africa. More than thirteen thousand houses have been built on that continent as of early 1999. And thousands more will be built there in the years to come. Harry Goodall, the son of Disciples missionaries to the Congo in the sixties and a very talented and dedicated man, is directing our growing work in Africa.

As mentioned in chapter 3, in January 1998, Linda and I and our daughter Faith visited our work in three African countries with Harry Goodall. In Tanzania, we saw hundreds of houses that had been built in Kasulu and other locations, primarily in the northwestern part of the country. In Zambia, we visited our work in three locations, all near the capital of Lusaka. In the Kafue Flats, we saw scores of tidy houses and stayed overnight with one of the happy homeowners.

In Ghana, we were amazed by the dynamism of the program. Among other things, we dedicated the one thousandth house built by Habitat for Humanity in that country. Nana Prah, a member of the Habitat International board, is one of the driving forces behind the work in Ghana.

In all of the countries, there were huge outpourings of joy and celebration. People were happy to partner with a ministry that enables families to have a decent home on terms they can pay. Faith captured the highlights of our trip on videotape. She produced a spectacular record of the trip and made it available to local affiliates that support the work in those countries and to others who had an interest in Habitat's work there.

In South Africa, Habitat for Humanity got off to a slow start, but things are now turning around in that country. Kurt Firnhaber, the international partner who has been regional director since 1996, is doing an incredible job of "putting Habitat on the map" in South Africa. The enclave of Alexandra in Johannesburg is where we first started building in that country. Nineteen houses were built before the construction stopped because of the violence. Now work has resumed, and it is expected that a great number of houses will be built there for years to come.

Faith Fuller videos her father spading mortar onto another house going up in Chanyanya, Zambia. Photo by Linda Fuller.

Just north of Johannesburg is a place called Orange River Farm. More than forty houses have been built there, primarily for a group of widows who lost their husbands in the recent violence. Yet another dynamic Habitat affiliate in South Africa is near Durban, in a place called Pisang River.

Other affiliates are being formed in South Africa, including some in the Capetown area. The need for housing is enormous in South Africa, and Habitat for Humanity intends to be a significant part of filling that huge need.

The most Habitat houses that have been built in Africa are in the small country of Malawi, in the southeastern part of the continent. Linda and I were there for a visit in 1995. We were impressed with the neat houses built in numerous towns and also in many rural areas. Habitat is building in one hundred locations in Malawi and, as of the end of 1998, a total of more than forty-five hundred houses had been built. More houses are going up in Malawi at the rate of about three hundred and fifty a year.

Come with Us to Asia and the South Pacific

And how about the rest of the world? Habitat for Humanity is flourishing in Asia and the South Pacific. The work started in the Philippines and India in 1985 and has spread gloriously into the neighboring countries, including Thailand, Sri Lanka, Korea, Japan, Nepal, Malaysia, Australia, New Zealand, the Solomon Islands, Papua New Guinea, Fiji, and Bangladesh.

In 1998, Steve Weir, director of Habitat's work in that part of the world, organized a trip for Linda and me to visit several countries. Let me take you with us and explain the state of our work in this interesting, beautiful, and needy part of the world.

Steve accompanied us, along with our daughter Faith, who documented this trip on video too. During a three-week period in July, we spent time in Sri Lanka, Malaysia, Singapore, Korea, Japan, and the Philippines. We were enthralled by everything we saw and experienced along the way.

Our first stop was Sri Lanka. Steve had actually started the work there along with his wife, Debra, in 1993, when they went there as International Partners with their children, Danielle and Kristin. By the time they left in 1995 for Steve to assume the position of Asia-Pacific director, work was underway in four locations and approximately twenty-five houses had been built. In the ensuing years, progress continued so that by the time of our visit there were five affiliates that had built four hundred houses.

We visited some prospective sites near the national capital of Colombo and one existing affiliate in that area. Then we drove to the interior, to the town of Hatton, in the middle of the tea-growing country. Linda and I spent the night with a homeowner family in Hatton and participated in dedicating a new Habitat house. We also had a big rally at the meeting center of a Catholic church. By early 1999, more than five hundred houses were occupied by Habitat families in Hatton and four other locations.

From Sri Lanka we flew to Kuching, Malaysia. How does a new effort in a new country begin? Sometimes by sheer will and excitement. John Chin, pastor of First Baptist Church in Kuching, spearheaded an effort to start the first Habitat affiliate in that country. We met there with nearly a hundred people who were wonderfully excited about bringing the ministry to their city and country. On Sunday, I preached in St. Faith's Anglican Church and at First Baptist Church. International partner Patrick Foley was also instrumental in launching the work in Malaysia. Due to his good work and that of John Chin and their Habitat partners, the work in Malaysia was officially approved by the Habitat International board of directors in October. We have a bright future in Malaysia.

The next stop was Singapore. Eagles Communication, a Christian organization that works with young people, and World Vision cosponsored a banquet at the Amara Hotel. The purpose of the event was to inform leaders in that city-state about the work of Habitat for Humanity and to explore the possibility of starting a Habitat support organization. Singapore has no poverty housing, but there are more than three hundred churches and many concerned and compassionate business and religious leaders who we thought would respond positively to the message and concepts of Habitat for Humanity. The day following the banquet attended by 150 people, we had a round-table discussion with pastors and church leaders. The future will reveal what role Singapore will have in the growing presence of Habitat for Humanity in that part of the world. The seeds we planted there will hopefully bear abundant fruit in the years to come.

From Singapore, we flew to Korea. Kun-Mo Chung, a dynamic member of the Habitat International board of directors and a well-known Christian and community leader in Korea, was key in organizing our itinerary in the country. Our first series of engagements was in the northeastern part of the country, an area that is very mountainous and beautiful but quite poor. The closing of many of the local coal mines brought on high unemployment and poverty. This is exactly the place that Habitat needs to be, and an affiliate was begun in Tae'beck City. A couple of houses had been completed and several more were under construction.

We visited both the completed houses and those under construction. We learned that more than seven hundred volunteers from Tae'beck City and elsewhere in Korea helped build the first houses. That incredible outpouring of support boosted the fledgling affiliate. The support was as imaginative as the country was needy. For instance, while in Tae'beck City, twenty-four bicyclists arrived from Seoul. They had been pedaling for four days. Their efforts netted the equivalent of three thousand dollars for the growing work of Habitat for Humanity in Korea.

That evening a concert was held in the city for the benefit of Korea Habitat for Humanity. In addition to some wonderful music, Kun-Mo Chung spoke and so did I. Twelve hundred people were in attendance. We returned to Seoul by train, spending the night in a sleeper car, an incredible experience for car-bound Americans.

Making a Habitat Member of the President of Korea

The following day I met with Korea's president, Kim Dae-jung at the Blue House, Korea's presidential palace. With me for the meeting was Dr. Chung, Wong-In Koh, a longtime leader in Habitat's work in Korea, and Linda.

The president graciously received me but was very formal in his demeanor. I spoke first, telling him about the worldwide work of Habitat for Humanity and explaining the philosophy of the program. I then talked about the growing work in Korea, explaining that building was underway in two locations, Tae'beck City and Ui-Bong-ju. and that new affiliates were being formed in two more locations. In addition, I mentioned Jimmy Carter's involvement in the work. He has a special respect and appreciation for former President Carter because while he was in office, he was instrumental in rescuing President Kim, then an opposition leader to a previous government, from a life-threatening situation.

Following my remarks, he spoke, but always in Korean. A translator sat immediately behind him and rendered his remarks to me in English. He spoke very favorably about me, President Carter, and the ministry of Habitat for Humanity. At the end of his speech, I invited him to become an official member of Korea Habitat for Humanity, and he quickly agreed. Dr. Chung brought the form to him and he signed it on the spot.

I told President Kim Dae-jung about Korea Habitat's plan to be building in a hundred locations within five years. He pledged his support and the support of his government for this growing movement in his country.

The next evening five hundred people attended a banquet at the Hilton Hotel, including a large portion of the diplomatic corps in Seoul. There was delightful musical entertainment and several speeches, including a video greeting from the first lady of Korea, Hee-Ho Lee. As in Tae'beck City, I also spoke, giving encouragement for the growing work of Habitat for Humanity in Korea. Then on Sunday I preached at two different churches, one a Presbyterian church of normal size and also at Kum-Ran Methodist Church in Seoul. This eighty-five-thousand-member church is the largest Methodist church in the world. Each Sunday there are five services, each attended by a packed house of more than six thousand people. After I spoke, the pastor, Rev. Hong-Do Kim, presented me a gift of the equivalent of ten thousand dollars for use by Korea Habitat for Humanity. That afternoon we visited the Habitat homes built by Ui-Bong-ju Habitat for Humanity.

Incredibly Committed

The leaders of Korea Habitat for Humanity, including the dynamic national director, Young-Woo Choi, are committed not only to rapidly spreading the work to more and more towns and cities in Korea, but they are equally committed to tithing and supporting the work in other countries. They are also very active in recruiting volunteers to participate in Global Village work projects.

In 1997, Korea Habitat for Humanity launched a major partnership with Habitat for Humanity in the Philippines. This partnership resulted in raising sufficient money in Korea to build twenty-five houses at various affiliates in the Philippines. More than three hundred and fifty volunteers from Korea journeyed to the Philippines throughout the year. These volunteers came from universities, churches, corporations, and high schools. They returned to Korea with a greater sense of the Habitat vision. Most are now actively involved in building Habitat houses in Korea.

In August 1997, a volunteer team from Korea joined with a volunteer work group from Japan to work in the Philippines. This Habitat partnership provided an awesome example of how people who traditionally have been at odds can unite around a hammer and work *together* to bring hope to families in need of decent shelter.

In 1998, Korea Habitat continued its partnership with Philippines Habitat for Humanity and the partnership with the Japanese has been maintained. Korea Habitat is also expanding its partnership to other countries in Asia. Korea Habitat for Humanity is on the move, and so were we.

In Japan—"Habitat Had Changed Us"

The next country on our itinerary was Japan. We flew into Osaka and went from there by train to Kobe, where in the evening we had a service at the Kobe Union Church. The church was filled mostly by university students from the four campus chapters of Habitat for Humanity in Japan: Kwansei Gakuin University, Kyoto University of Foreign Studies, Kwansei Gakwin Uegahara University, and Dishisha University. The students were so excited. In fact, I have never seen more excited or motivated students as those Japanese students were about Habitat for Humanity. Two students spoke that evening, including a young woman, Mariko Asano from the Kyoto University of Foreign Studies chapter. She captured the hearts of everyone with her moving account of her involvement in Habitat for Humanity, especially her trip to the Philippines to build Habitat houses. She talked about a woman named Sarah who she met in a decrepit shack built over a dirty lagoon. Then she told of working with Sarah on her new Habitat house and of the thrill of being a part of changing the life of that woman and her family, particularly Sarah's unborn child.

Mariko concluded, "The Habitat experience is big, personal, and emotional. We are not so sure just what it means to us, but it feels right. It pulls you together somehow. You get a feeling you are going in a new direction. It is hard for us to express more clearly, but it is not a feeling of confusion or uncertainty. We realized that Habitat had changed us. We had learned something that was not going to be unlearned. We are not satisfied with drifting, with accepting things as they are. We don't want to change and pick up our old lives again. Our new direction is forward."

The first campus chapter of Habitat for Humanity Japan was started by professor Akiie Ninomiya, a dynamic Christian leader at Kwansei Gakuin University in Sanda City, Hyogo-Ken. Within a week of announcing the start of the chapter in 1997, fifteen students were meeting weekly in Professor Ninomiya's office. Within four months, the chapter was officially approved by the Campus Chapter and Youth Programs department of Habitat for Humanity International, and twenty-five students were off on the first of many volunteer trips to work with Habitat affiliates in the Philippines. By the end of 1998, the Kwansei Gakuin chapter and the three other Japanese chapters had sent seven volunteer teams to the Philippines and one team to the United States. Nearly two hundred students have participated in these service opportunities and have raised funds for eight houses in the Philippines.

The students at the Kobe Union Church told me that they wanted to form Habitat Campus Chapters at every university in Japan. With their motivation and enthusiasm, they will probably do it.

After the wonderful meeting in Kobe with the students, we took a train the next day to Tokyo. Mariko Asano and her friend and Habitat coworker, Mari Sano traveled with us. That night we had another fine meeting at Tokyo Union Church, which was attended by some students but also by many other people from many walks of life.

Philippines—Ambassadors, Presidents, and Wrong Way Streets

The next day we headed south for the last country on our Asia tour—the Philippines. Linda and I had last been in the Philippines in 1993. On that visit, we dedicated the five-hundredth house built in that country. On this visit, we would dedicate the two-thousandth house. We would also help prepare for the big Jimmy Carter work project to be held there in March 1999, which was scheduled to build two hundred and fifty houses in a week in six locations throughout the country. (In fact, a total of two hundred and ninety-three houses were built that week!)

A portion of the 130 houses under construction at the Maragondon site during the March 1999 Jimmy Carter blitz-build in the Philippines. Photo by Gregg Pachkowski.

Dedicated Habitat leaders Gloria Ison, Andrew Regalado, and others met us at the Manila airport. We were then led into the city by police escort.

In the following three days we visited the U.S. Ambassador, Thomas Hubbard, talked to numerous Habitat leaders, and I addressed the U.S. Chamber of Commerce of the Philippines, the Manila Rotary Club, the oldest and largest Rotary Club in Asia, made a talk at the Union Church of Manila, and spoke at a big banquet of Habitat supporters. I also had a personal meeting with the new president of the Philippines, Joseph Estrada. He told me and those in our group that his meeting with us was the best thing that had happened to him in his first thirty-one days in office.

President Estrada was very outgoing and friendly, in sharp contrast to the stiff formality of Korean president Kim Dae-jung. He promised to work with President Carter during the blitz-build in March.

At several points during our stay in the Philippines, we had police escorts. These escorts, with sirens blowing and lights flashing, would often go down one-way streets the wrong way because traffic on the other side of the street was stalled. It was exciting but a bit unnerving! We did move a lot faster through the incredibly crowded streets of that capital city, however.

Habitat International board member Sonny de los Reyes presided at a press conference where the announcement was made about the build with President Carter. Our last event in the Philippines was to dedicate the two-thousandth house. It was in Topsville, a community that was just getting started when Linda and I visited the Philippines on our first visit five years earlier. We saw the first half-dozen houses under construction on that visit. Now, a total of one hundred and forty-five houses stood complete and occupied. Only six remained to be built before the entire development was finished. We learned that the community had recently won an award for being the best neighborhood in that part of Manila. The Philippines has numerous Habitat neighborhoods that are thriving, including Rotaryville and Kamarin.

Bedecked with Flowers—
Two Perfect International Neighborhoods

In Negros Oriental, in Dumaguete City, Habitat has built two big communities: Balugo with eighty-eight houses and Candau-ay with eighty-five. When Linda and I visited there in 1993, we were thrilled with what we saw. We agreed that those were the most beautiful Habitat developments we had seen anywhere in the world. Each house had flowers growing in profusion. Neat little gardens were across from narrow paths from the houses. Painted rocks lined the paths in front of the houses.

The government had built a school for the children. Dedicated Habitat partner Tom Dineen of Foxboro, Massachusetts, had provided funding for a lovely community building and play area, and he had financed a water tower for the people.

Many of these families had previously been living in an adjacent garbage dump. The government closed the dump and turned it into an ecosystem park, planted with trees and a bird sanctuary. More than houses? That "more" is certainly what happened in Dumaguete City.

Self-Sufficient Community of Pride

Another wonderful example of "more than houses" is in General Santos City in southern Mindinao Province. There Habitat for Humanity has built more than three hundred houses as of mid-1999, and the transformations have been extensive and truly exciting.

In 1987, General Santos City Habitat Community Association was formed to acquire land. The association bought thirty-six hectares (ninety acres) on which the three hundred–plus houses have been built. Today the association has its own system for conflict resolution, including a people's court for cases that cannot be resolved through other means.

The community generates income to support its programs. They operate a concrete block "mini-factory" that makes and sells blocks. This enterprise also provides jobs for some Habitat homeowners. Part of the land is under contract with Dole Philippines for a banana plantation, which generates more income for the association. Also, as with the concrete block enterprise, some homeowners work for Dole.

The association assists homeowners in making and selling handicrafts and also provides financing for members to start their own small businesses. The association also provides water to the homeowners by two deep wells, one of which was financed by the association and the other through the local congressional office. Lines for electricity were installed and paid for by the association too. Homeowners pay the local electric cooperative for connections and monthly usage.

There are two nursery and kindergarten schools in the community. The government operates one of them and the other is private. The community initiated the private school. The facilities for the school are provided free of charge, and the teacher is paid by the association.

Beyond the extensive operations outlined above, the association plans to extend its services to more people. It intends to acquire additional land so that more needy families can have a decent house in a good community. Without a doubt, the enterprising people of General Santos City are building houses and a *whole lot more*.

South America, Too—"I Knew Religion, Now I Know Love"

Half a world away in the country of Peru in South America, Habitat for Humanity is also building a lot of houses and a whole lot more. Andy and Debbie Kramer are two of our most dedicated and effective international partners. They have served both in Peru and Mexico. In the mid-1980s, they moved with their children—Eric, Wesley, and Jesse (Kristin was born in Peru)—to Juliaca, a town in the high Andes in southern Peru, near Lake Titicaca. Habitat had just acquired a parcel of land at the edge of the city large enough for more than three hundred houses.

A group of twenty-four families was chosen to work together to build their houses on that land. In the group was a young, vibrant, outgoing, handsome man named Marciano. He was chosen to be the spokesman for the families. Marciano, his wife, and their young daughter were faithful in participating in all aspects of the program, including the morning devotions. They were always there. The devotions were made up of reading a few passages from the Bible, praying, singing, and hearing a short message, which was always simple and nondoctrinal in keeping with Habitat practice. Finally, one morning after the brief devotional, Marciano rose to his feet and, with a smile on his face, said that while he appreciated the Bible, they already had their Catholic religion, and what they had come to Habitat for was a much-needed home.

The devotions and sharing continued and so did Marciano. More than fifty people would come to build each day. Women with babies on their backs would be laying bricks while chatting away. Men, often singing and laughing, were digging foundations or nailing together roof trusses. They would often end the day with a soccer match. Always, children were everywhere, helping carry water and bricks or just playing marbles in the sand or making dolls from the tall, stiff pampas grass.

When the twenty-four houses were finally finished, all gathered for a joyous celebration at which the houses were dedicated and a Bible presented to each family along with the keys to their new home. The jubilant group moved from house to house, and in keeping with local custom, bottles of champagne were broken over the doorways. When the group arrived at Marciano's house, and he was presented with his new Bible, he just stood there for a long awkward moment while tears ran down his cheeks. Finally, his face beaming and clutching his Bible with both hands, Marciano said, "When I first came out to work here, I knew religion. But now I know love. I've learned so much from all of you together and from this book. How can I ever be the same? And how can I ever thank you?"

Not Houses but Community

In 1988, after serving for two years in Juliaca, the Kramers were given a party by a group of the Habitat homeowners as they were leaving to go to another city to start a new Habitat affiliate. That occasion was another powerful testimony as to how Habitat for Humanity is building so much more than houses. Andy remembers what that occasion was like: "One by one, several people stood to speak, but much to our surprise not one mentioned anything about their new houses. Instead, each spoke of their new community.

"I remember one man in particular, Cipriano Blanco, whose hands began to tremble as he stood. He haltingly told us of how much he'd valued seeing for the first time what a Christian community, a Christian family, and a Christian marriage could truly be. It was an example, he said, that would always remain with him and that he had vowed to live up to. Tears streaked his face as he sat down. We were never so aware as at that moment how much more than just houses were being formed through the ministry of Habitat for Humanity."

Moonscape Transformation

The next assignment for the Kramers was the city of Arequipa, some two hundred and thirty miles north of Juliaca. When they saw the land that had been given for the Habitat houses, they were shocked. Andy described the property: "It was like a moonscape, located far from the edge of the city, covered with sand and cinder ash on the barren slopes by El Misti, the volcanic cone that rises twelve thousand feet over the city." Andy wondered if people would want to live there. Besides that, would they even be able to get out to the site?

The land was laid out on paper to include more than six hundred Habitat houses plus areas for parks, schools, and churches. The local Habitat for Humanity affiliate had already named the land El Nazereno. Lots would be given free of charge to any church that wanted to build there. Word was spread throughout the city of Arequipa about the free land for churches. And even though the city had a million people and in excess of a hundred churches, only one man responded to the offer. His name was Willy Bustinza.

Andy said that he was a quiet, humble pastor of a tiny evangelical church in one of the poor barrios of Arequipa. Willy and his wife, Aida, had dedicated their lives to working with the poor and marginalized people of Peru, so the challenge and opportunity from Habitat was very appealing to them.

Willy and Aida started coming to the new site to assist in the building of the houses. They also joined the homeowner families in daily devotions and on Sundays held a small Sunday School for the many children who came out with their parents to work. Sundays were often the only days many of the people were free to help with the construction.

Soon, Willy and Aida and friends laid the foundation for a small two-room church and school building. They named the building El Pionero.

By the end of the first year, 1988, several families had moved into their new Habitat houses, and Willy's church and school were built. Worship services were held in the little church on weekends, and eight children started classes at the school using concrete blocks for chairs and pieces of torn cement bags to write on. Realizing the value of the local school, many of the new homeowner families helped add on more classrooms as the number of children in El Nazereno steadily increased.

Andy and Debbie returned to the states with their children, but their hearts remained in Peru. When they received an unexpected inheritance, they made a special gift to build more classrooms and hire two more teachers.

Willy's School

Over the next few years, the school continued to grow as more and more children moved into the community. A hundred new houses were being built every year in El Nazereno. Willy and Aida helped organize a small health clinic and staffed it with volunteer professionals from Arequipa.

As the school continued to grow, the need for teachers naturally escalated. Willy refused to charge students for the costs of the school. Instead, he turned to the government. Peruvian law, historically, had provided funding for teachers in private Catholic schools. Protestants, however, were not extended the same privilege. Times, however, were changing in Peru, and the combined efforts of Willy, Aida, and others brought about a change in the law so that assistance could be given to all faiths. Willy received his requested help from the government for more teachers.

Enrollment at the school doubled each year with the enthusiastic support from the growing Habitat community. Although more than 90 percent of the affiliate families were Catholic, most appreciated Willy's dedication, sincerity, and simple, practical faith. Willy didn't preach against Catholicism, as did many Protestant churches in Peru. Rather, he preached and practiced God's love and grace. This was abundantly clear to everyone, parents and students alike.

A Christian foundation in Norway learned about Willy and Aida's work and made a commitment to provide funding for the construction of new school buildings. Following Habitat's example, several work camps came over from Europe to assist with construction, while a few North American donors pledged money to cover the administrative costs.

By the end of the first decade, El Pionero had more than eight hundred and fifty students in classes from preschool through high school. Several former students have gone on to become teachers, pastors, and social workers.

The demand for more classrooms in El Nazereno continues, and the demand for more schools in other locations also continues. A neighboring community has formed a committee requesting that El Pionero build a school in its neighborhood. The committee has acquired a plot of land and is ready to provide all the labor needed for construction.

The Habitat families of El Nazereno have faithfully volunteered their time to maintain and expand the school. The church has grown as well and is a visible sign of God's love in the community.

By December 1996, 588 families lived in El Nazereno. No more Habitat houses could be built there because all of the land was utilized. More lots had been used for the school and for other community purposes. Other Habitat houses were being built, though, in locations elsewhere.

Catalyst for Other Development

The Habitat community at El Nazereno has proven to be the catalyst for a surge of development in the surrounding area. The local Peruvian government built nearly two thousand tiny but solid "starter houses" for its employees along one side of El Nazereno. Other families purchased small lots and built homes on their own. The barren "moonscape" has been dramatically altered in the short space of less than a decade.

In writing about El Nazereno, Andy Kramer concluded, "Habitat for Humanity has made a difference, providing a community through which God's promise and presence have become real and lasting." More than houses? In El Nazereno that is certainly true.

In eighteen locations in Peru, Habitat communities are growing into reality. I've visited many of those communities on two trips to the country, including Puno, where we built El Mirador, a neighborhood of 163 houses on a high plateau overlooking the northern shore of Lake Titicaca. Puno was the first place we built in Peru. I've also visited Tacna near the border of Chile in the far south of the country where more than 530 Habitat houses have gone up. I have promised to return to Peru to dedicate the five-thousandth Habitat house built in that country. As of early 1999, they are past four thousand houses, so I'll be headed south for Peru very soon.

Elsewhere in South America, Habitat for Humanity is also making a huge difference. Many hundreds of houses have been built in Bolivia, Colombia, Guyana, and Brazil. And work is being launched in Ecuador and Paraguay.

In Central America, Habitat is building and transforming communities in all countries except Panama. We also have extensive work in the Caribbean in a growing number of countries.

In all of these countries, solid houses are being built, but we build more than houses. That is always the aim of this ministry—to build houses and strengthen individual families, but also to build and develop communities.

Largest Work Outside the United States—Mexico

Our largest work outside the United States is in Mexico. More than ten thousand houses have been built in that country. And we continue to build more than a thousand houses every year. Again, the emphasis is on community development along with the building of individual houses.

Andy Kramer wrote about the "more than houses" aspect of the work in Mexico. He and Debbie came back to the United States from Peru in the late eighties. They returned as international partners to Mexico in 1991. The place they served in Mexico was Huasteca, in east central Mexico. That affiliate encompasses some thirty villages. Andy describes the area: "Undisturbed, overgrown mounds—the ruins of an ancient culture were scattered throughout the region. Bits of pottery and arrowheads are common in this area where many people still speak their ancestral language of Nauhaut."

Andy praised the local Habitat leaders, José Hernandez and Justino Hernandez, saying both are tireless, committed workers, even though they each have only a first-grade education.

When the Kramers left the Huasteca affiliate in 1992, forty houses had been completed. Since then, another four hundred have been built.

Too Great Not to Be Shared

But one of the great contributions of that affiliate, beyond building houses, has been the *leadership* it has demonstrated for other areas of Mexico. Both José and Justino were accepted as national partners to lead different affiliates in the state of Chiapas. Two others, both homeowners, were chosen to head affiliates in Veracruz and in the northern mountains of the Sierra Madre. "For these people, the hope offered through Habitat was too great to be contained within their own communities," Andy said. "It needed to be shared."

Hurricane Mitch's Severe Test—Passed!

All of the exciting work in Mexico and throughout Central and South America and the Caribbean is headed by Steve Mickler. He and his dedicated and able staff completed a move to Costa Rica in July 1998. From that location they are better positioned to build tens of thousands of houses in the years ahead and, as Andy and Debbie Kramer report, so much more. A severe test came to Steve and his staff and to all of us in leadership of Habitat for Humanity International in October 1998 when Hurricane Mitch roared through Central America, leaving death and almost indescribable destruction in its wake. Thousands of people died in Nicaragua, Honduras, Guatemala, and El Salvador, and tens of thousands of houses were destroyed. We literally held our breath wondering how our Habitat houses had fared. But we need not have worried. Only a few of the more than sixty-five hundred houses we had built in those countries were seriously damaged. But thousands of houses in the various areas of our work were destroyed. So we swung into action, and within five days we were putting five hundred thousand letters in the mail asking for special contributions for an emergency building fund. By year's end that fund held more than five million dollars, including a one million dollar grant from the John S. and James L. Knight Foundation —enough to build another two thousand houses. When this book went to press, those houses were rapidly going up!

Even in Europe

In a very different part of the world, namely Europe, Habitat for Humanity is a growing presence. Habitat for Humanity of the Netherlands, as we mentioned earlier, has a fine talent for raising money and recruiting volunteers for work in other countries, and we hope to launch other such Habitat organizations in other Western European countries. Habitat is already building houses and neighborhoods in a growing number of countries in both Western and Eastern Europe, including Poland, Hungary, and the United Kingdom.

In the United Kingdom, the first two houses were built in Banbury in 1997 and then four Habitat houses were built in London in 1998 and more in Banbury. In Northern Ireland, under the inspired leadership of Peter Farquharson, eleven Habitat houses have been completed in the Catholic neighborhood of Iris Close Estates and six more put up in Glencairn, a Protestant community.

Crossing the Peace Line in Ireland

The "more than houses" that is taking place in Northern Ireland is *reconciliation*. Catholics and Protestants *are* working together. And that collaboration is growing month by month.

Special events highlight the work in that beleaguered part of the world. Numerous work groups have gone to Belfast over the past few years and more are being planned all the time. One very special work group was Blitz-Build '98. Four houses were completely built in Glencairn in two weeks. The blitz started on May 18 and concluded with a moving dedication ceremony on May 30. More than three hundred and fifty local volunteers, Catholic and Protestant, joined thirty international volunteers, including a work team of Habitat International board members and the four homeowner families to build those four houses.

On May 20, U.S. Ambassador to the Court of Saint James's, Philip Lader, was a volunteer. He said about the occasion, "I am pleased to support the work of Habitat for Humanity in Belfast. Not only will this project help provide

Linda LeSourd Lader, wife of U.S. Ambassador Philip Lader, on a Habitat build site in Southwark, London, in May 1998. Photo by Robert Baker.

much-needed housing for all segments of the community; it will also contribute to the economic regeneration that is vital to the establishment of lasting peace in Northern Ireland." Another very special visitor to the site was the U.S. Ambassador to the Republic of Ireland, Jean Kennedy Smith.

Peter Farquharson said that Blitz-Build '98 was a tremendous success not only in terms of the building program, but most importantly, from the community development aspect of the work. "The Iris Close Catholic families took a step of faith, crossed the peace line, and worked alongside the Glencairn Protestant homeowners during the blitz," he continued. "Now, the Glencairn Habitat families are proud homeowners and trust is being nurtured between the two Habitat communities. One local clergy said that he felt this work has really affected the local churches who 'saw the Gospel in action and smiles on faces in the community.'"

Peter now plans to make Belfast Habitat a regional organization to respond to growing interest in the ministry in places like Downpatrick, Lisburn, Newry, Portadown, and other locations both in and outside Belfast. He and his dedicated coworkers are also excited about their millennium project, which is the possibility of an *integrated housing development*—Protestant and Catholic families *together*—on the Springfield Road in Belfast. Peter says this project would be the realization of their vision and mission for the work of Habitat in Northern Ireland and could serve as a radical model of transforming and uniting the whole community.

Brick by Brick—Reconciling

Habitat for Humanity International board member Paul Leonard participated in Blitz-Build '98. As he expressed it, "We are building simple, decent houses for people in need, but this is not the only reason we are in Belfast. Brick by brick we are building community and tearing down the 'peace walls' so that the peace that comes from God can reconcile our Protestant and Catholic brothers. Our blitz-build was a symbol of that reconciliation. Protestants and Catholics worked side by side, not caring about our individual traditions, but sharing our love of the one God who called us into being and who commanded that we love one another. Using our hammers, we were acting out that love in partnership with the families who would soon own and occupy their new homes."

Hungary—The Surprise of Volunteers

Quite a different kind of community was built in Vac, Hungary in August 1996 when former President Carter, his wife, Rosalynn, and five hundred volunteers from twenty-five nations descended on that small town on the east bank of the Danube, some twenty miles north of Budapest, to build ten houses in a week.

The local people had never seen anything like it. Their greatest surprise was learning that all those people came at their own expense to work as *volunteers*. But they were also surprised to see the houses built so quickly and to realize that they were strong and well built.

The blitz-build there solidly launched the work of Habitat for Humanity in Hungary. Since then, sixteen more Habitat houses have been built in Felsogod, a town between Vac and Budapest, and four more have been built in Csepel, a suburban community of Budapest. The blitz also had the effect of spreading the word about the ministry to other countries in Europe. As this book goes to press, new Habitat organizations have been formed or are forming in Slovakia, Germany, Bulgaria, and Portugal.

One country that sent a large delegation to Hungary was Romania. Habitat for Humanity had been organized in that country under the leadership of a very dedicated Christian man from Beius named Adrian Ciorna, but no houses had been built. Since that time, seven houses have been built in Beius, and efforts are underway to launch building in Arad.

Engine for Community Development

Adrian told about the beginning of the work in Beius and about how Habitat's work precipitated many other good things:

> The local government donated eighteen hundred square meters of land to Habitat for Humanity, enough land for eight houses. It was a totally undeveloped area adjacent to another thirty thousand square meters of land that had not yet been used for building. Some other families had been given this land in 1990, right after the Romanian Revolution of 1989, but no one had built anything. There was one family of nine, we discovered, living on the land in an old hog pen! I was told during my Americus training that Habitat is an "engine" for development in the community. I didn't realize the full dimension of that statement at the time. Now, I do.

A couple of months after we started building in April 1997, three more houses were erected by three other families. A dozen more families planned to start building soon. They believed that if Habitat was there building, infrastructure would come soon.

When we started to clean a ditch to evacuate rainwater, two other neighbors did the same. Habitat for Humanity of Beius applied for electricity for two houses to be built in 1997. And even though it was most unusual, electricity was provided not just for the two houses, but for all eight Habitat houses and houses to be built on five other streets.

People view very highly what is going on in the neighborhood and appreciated the quality of the houses. The passive solar design and the way the space in the house is divided inspire many around town to transform other neighborhoods.

The extended family in the hog pen decided to improve their humble "house" (two rooms—one hundred and forty square feet and a hall) by installing a new door, fixing the window, and painting one of the rooms. Habitat helped them with some limestone and cement.

These are a few extraordinary ways in which the local Habitat affiliate changed for the good the neighborhood where it started work. I look forward to many other rewarding experiences, through God's grace.

Eclipse Build in Romania

From August 1 to August 14, 1999, Habitat for Humanity Romania will work with hundreds of volunteers from around the globe to build ten homes in partnership with Romanian families in need of decent shelter. This event will be called the "Eclipse Build" because it will happen during the final total solar eclipse of the millennium. An Englishman, Ian Walkden, is the director of our work in Europe. He and his able staff, especially Doug Dahlgren, working out of Hungary, are working diligently to expand the ministry in that part of the world and also to Russia and the newly independent states.

Ian Hay, well-known homebuilder and currently a Habitat for Humanity International board member, started Habitat for Humanity in New Zealand in 1993. He has expanded the work in that country to twenty towns and cities. Ian and his Habitat partners have planned a special midnight blitz-build at the end of 1999 to usher in the new millennium.

As everywhere else in the world, we continue to build houses and *so much more*. To touch the world and change it, to reach beyond one's own culture with the practical example of love in the mortar joints that Habitat practices, to offer all of God's children—and we all are God's children—wherever they may live, the foundation of a better life through a decent home in a decent community is worth far more than the two thousand dollars it may cost. Is Habitat about more than houses to people beyond our borders? Just ask Ian, Adrian, José and Justino, Willy and Aida, Marciano, Mariko, Young-Woo, and all the neighbors who live in Tae'beck City, El Nazereno, Iris Close Estates, flower-bedecked Balugo or Losanganya, the "place of reconciliation."

They would probably smile and ask you to come in.

CHAPTER 6

Improving Health

*You are doing more for the health of the people in this village
than we have with our clinic in twenty-five years. You are
getting rid of the reasons for their illnesses.*
—BRITISH BAPTIST NURSE IN ZAIRE

Many things naturally go together—husband and wife, open and
honest, sweetness and light, neat and clean, robust and healthy.
Good housing and good health also go together. They are entirely comple-
mentary. One begets the other in surprising commonsense ways. A well-
built house with solid walls, a good roof, and other simple, basic features
promotes health in powerful ways. A safe, secure house enables the occupants to sleep peacefully at night. Adequate rest promotes healthy living. A
sense of security that comes from good doors and windows that can be
locked gives peace of mind.

But most of all, a simple, decent place to live, built solid and true, makes
cleanliness possible. A solid roof, essential to any house that can be called
"decent," keeps the rain out, an enormous boost to good health, and solid
walls keep out rats and mice, along with other vermin.

Dealing with the Reasons for Illness

That may all seem obvious to us, but many people cope with rain and rats
and vermin and dirt and insecure doors every day of their lives. Many poor
families, especially in developing countries, live in houses that leak, have
dirt or cow-dung floors, and have no toilet and no suitable place to take
a bath. Such living conditions are conducive to sickness, stress, and pre-
mature death.

I remember a conversation I had years ago in Africa with a British Baptist nurse in Ntondo, Zaire, about our burgeoning work there. "You are doing more for the health of the people in this village," she said, "than we have with our clinic in twenty-five years. You are getting rid of the reasons for their illnesses."

Wall Came Crashing Down

I have heard and seen these reasons. I know the feeling, the fear. I have slept under such roofs and listened to the sounds that kept me from sleep and from safety. It isn't an experience anyone would desire.

While I was traveling in the interior of Zaire in 1975, my party spent the night in a remote village. I was sleeping on a bamboo-cane cot in a mud hut. In the middle of the night, a fierce rainstorm hit the village. I lay there awake as the rain crashed down on the fragile hut with increasing ferocity. I heard sounds in the wall that didn't sound reassuring. After a few minutes, the wall disintegrated and came crashing down. I got soaked with water and splashed with mud. Needless to say, my sleep was over. It was a sobering and unforgettable experience. I gained new understanding and appreciation for every house we build.

When I was in Zambia in 1998, a new homeowner exclaimed to me, "I'm so happy to be in my new Habitat house. Termites no longer get into my ears at night." It takes no imagination to understand that his new Habitat house has made him a happier and healthier man.

Rural Health Statistics

Statistics from the Chanyanya Rural Health Centre management team in Zambia reveal that Habitat houses apparently have made quite a difference in the health of the people in the Chanyanya area. Four hundred Habitat houses have been built in that rural community fifty miles southwest of Lusaka, the capital of Zambia. Each house has a pit latrine. We saw many of the houses during our 1998 visit. One night we stayed with a family living in one of the houses.

C. Chaila Cog of the Rural Health Centre team gave us the latest statistics for malaria and dysentery in Chanyanya:

DYSENTERY

1995: 361 cases	1996: 263 cases	1997: 48 cases

MALARIA

1995: 18,782 cases   1996: 15,644 cases   1997: 11,429 cases

Mr. Cog concluded, "Analysis of the statistical figures drive us as health staff and management to recommend your work and request extension to more communities in the catchment area."

Bishop Zebedee Masereka of South Rwenzori diocese in western Uganda reports that the hundreds of Habitat houses that have been built in Kasese and throughout the region have powerfully impacted the health of the families. "I do give thanks to God for what Habitat for Humanity has done in the lives of families that own Habitat houses in South Rwenzori diocese," the bishop exclaimed. "Children born in these houses are now big and living a happy life. Some families that were almost broken are now united and living happily together."

Solar Systems, Sanitation Facilities, Trees for Fresh Air

"Habitat in many ways has improved the living standard of the people, for example, through the introduction of solar systems," the bishop continued. "This has rescued pupils living in these houses from the burden of using poor lighting. Schoolchildren can now do their homework under proper lights.

Surrounded by raw sewage and trash, a mother bathes her child in Paranaque (Manila), the Philippines. Photo by Lloyd Troyer.

"People who used to stay in dilapidated, shabby houses are now enjoying the clean atmosphere of the Habitat shelters," he said. "Disease like skin rash due to bedbugs is now forgotten. With the sanitation facilities in place, worms are now rare cases. And the trees that have been planted give pleasant, fresh air each day and help control the environment."

Liberated from Mosquitoes, Worms, Fleas, and Cold

One of the families benefiting from a Habitat house such as the one Bishop Masereka described is that of Amos and Saloai Kaahwa and their children. The Kaahwas moved into their new home in the village of Kijwara, Uganda, in 1996. Saloai remembers what life was like in their old mud-daub hut. "When we used to sleep there, I kept the baby cradled close to myself to keep it from being bitten by rats." She said the hut was infested by termites too, causing the structure to be in danger of collapsing.

Robin and Doral Chenoweth, dedicated Habitat volunteers from Columbus, Ohio, traveled to Uganda in 1997. They visited the Kaahwa family at their new Habitat house. While there, a rainstorm came up. "Everyone retreated to the new house," Robin said, "where the raindrops beat a frenzied cadence on the metal roof." He also described the inside of the house: "All was dry, clean, and comfortable. Blankets and clothes were neatly folded on beds in two rooms. In the main room, Saloai sat on a low stool and watched, through an open window, as the rain poured outside."

From that safe perch, Saloai reflected on what life was like in their new home. "Here, we're not attacked by mosquitoes and the coldness from the holes in the walls of the other house. We're more warm and protected. When we got a house, we felt liberated. We had been sleeping in a leaking house full of dust, full of fleas. Now these things are no longer a problem."

Jason Phiri, his wife, daughter, and two grandchildren were the first people to move into a Habitat house in the village of Kasungu, Malawi, in southern Africa. And their move came not a moment too soon. A few days later, their old rammed-earth house disintegrated and fell down.

Tilly Grey, international partner working in Africa, reported that countless thousands of other families are languishing in pitiful houses throughout Malawi. Many, especially in rural areas, live in houses with damp earthen floors, insects in the roof thatch, and snakes and rats that cannot be kept out of the mud houses. Such houses create huge health problems, including the danger of collapse, as with the old Phiri house, with possible injury or death to the inhabitants.

Poverty housing in the Manila area. Photo by Kim MacDonald.

Rollins School of Public Health Study

In 1997 the Rollins School of Public Health of Emory University sponsored a study to evaluate the health impact of Habitat houses in northern Malawi. Specifically, the study was designed to evaluate the impact of better housing on the health of children less than five years old. The research and study were carried out by Christopher G. Wolff, Dirk G. Schroeder, and Mark Young in collaboration with the Ekwendeni Mission Hospital and Habitat for Humanity International in the northern Malawi town of Ekwendeni.

Improved housing subjects for the survey were randomly selected from a master list of the three hundred homes built by Habitat for Humanity in the area. After the houses were selected, a guide from Habitat for Humanity assisted in finding the correct household. The survey was then administered by a trained surveyor. Permission was also sought to collect a water sample from the household storage container and water source, collect blood from the finger of any child five years old or younger living in the house, and perform a medical examination of the child that included checking the spleen for signs of infection. After this, the closest non-Habitat house was selected, and the procedure was repeated.

Data from the two surveys performed in March and August 1997 were compared. The analytical data set consisted of 211 households with 318 children, of which 143 lived in 98 Habitat houses and 175 children under the age of five in the 113 non-Habitat houses.

The sample of Habitat and non-Habitat houses showed no significant differences in socioeconomic variables between the two groups. Both groups had an average of three rooms and six people per house. That's where the similarities stop, though. Significant differences were found in household characteristics between the two housing groups. Almost all of the Habitat houses were made of burnt bricks (94.9 percent), tile roofs (99.0 percent), and cement floors (96.9 percent). In contrast, non-Habitat houses had thatch roofs (91.8 percent), mud floors (90 percent), and walls of either unburnt bricks (43.8 percent), mud (26.8 percent), or burnt bricks (26.8 percent).

What was the biggest difference? The study found that children under the age of five years living in Habitat houses experienced 44 percent less illness than children under the age of five living in non-Habitat houses. That's an amazing difference! The authors of the study concluded:

Measuring the effect that housing quality contributes to health is complicated by a wide range of factors. Although the study has tried to incorporate as many of these factors into the analysis, physical housing structure and other tangible changes in the environment are not the only benefit a resident receives from quality housing that could benefit their health. The health of residents in improved housing may benefit from an improved outlook on their situation due to the improvements in their housing. In fact, when respondents to the survey were asked to rank their physical, emotional, and overall quality of life on a scale of one to ten, the mean responses for Habitat houses was greater than the mean responses for non-Habitat houses in all three cases.

Regardless of what aspect of housing contributed to health, this study strongly suggests that the Habitat for Humanity housing program significantly reduced the burden of disease in children under five years of age.

Making It Even Healthier

The authors of the study identified a number of areas in which future Habitat houses could be improved to enhance the health of the occupants even more. I met Christopher Wolff and others involved with the study, along with other key Habitat officials, to discuss how to make the suggested improvements. They specifically felt that incorporating clean-water provisions, screening for the windows, hygiene education, and bed nets or ceilings would make an even more substantial improvement in the health of the people living in Habitat houses.

We are thrilled with the positive findings of the survey, but we want to implement the suggested changes to make the houses even better and more healthy. We also want to build more houses just as fast as we can.

Habitat for Humanity fully realizes the peril of the people. That's why the work is being accelerated in Malawi. As of the end of 1998, more than forty-five hundred houses had been completed and plans are to build at least three hundred and fifty houses in 1999 and even more in the years ahead.

Better to Die

In India, Habitat for Humanity is building thousands of houses and vastly improving the health of the families in the process. Anita Mellott, a native of India, now married to former international partner Jim Mellott and working at Habitat headquarters in Americus, described the impact of a new Habitat house on two Indian families in Bangalore.

Mr. Irudyraj, his wife, and their three children lived in a mud-walled, thatched-roof hut. He said he couldn't begin to tell about all the problems he had living in that hut. He exclaimed, "I used to think it would be better to die than to live in those conditions."

Mr. Irudyraj said they suffered during the rainy season when the rains flooded the hut, preventing them from sleeping on the mud floors. The rains also destroyed the few belongings they had. "During the rainy season, in order to protect my family, " he said, "I would send them to a friend's house that was not flooded. Once, I was so desperate, I even sent my wife and children to another state."

All of that changed when the family moved into their new Habitat house. "No matter where I am," he said, "I can always come home to a decent, peaceful home."

Muniamma, a widow, moved into her new Habitat house with her family in 1993. She moved from a one-room, thatched-roof, mud-walled hut, where she cooked on one side of the room over a makeshift wood stove and the family slept on the other side on the mud-caked floor.

"When it rained, water would come into the hut, so we could not sleep. We were constantly worried about termites destroying the structure. Every time there was a heavy breeze, mud from the walls would fall. If we were eating, mud would fall into the food. We had no proper sleeping, eating, or cooking facilities," she said. "I had no peace of mind in the hut. I constantly worried about our safety. I used to worry about snakes coming in and falling from the thatched roof. I thought that living in a hut would be my life. But now, through Habitat for Humanity, I have peace of mind and we live comfortably."

Her Habitat house is constructed from concrete blocks and has a reinforced concrete roof. It has a hall, a bedroom, and a kitchen.

No Kitchen, No Latrine

Another exciting story of the transformation of health comes from a family in the small village of Manali, a few miles from Nagercoil, the headquarters of Kanyakumari District in Tamil Nadu, India. The president of Kanya-kumari Habitat for Humanity, Father S. Joseph, explained the family's situation.

The head of the family is Mr. Esther Raj. He was a mason by profession. He and his wife, Thulasi Mary, have two daughters. The family was supported by Raj's hard labor. In 1995, Raj fell from the top of a building and was injured. All of the family's savings were spent for treatment.

"They were all living in a small thatched hut of eighty square feet," said Father Joseph. "There was no kitchen and no latrine. They led a miserable life there, especially in the rain and under the hot sun."

Kanyakumari Habitat for Humanity learned about the sad plight of the Raj family and chose them to have a Habitat house. When Raj heard that his family was going to have a new house, he was very happy and his health began to improve immediately.

The house was completed in November 1997. With an area of three hundred and sixty square feet, it is more than four times bigger than their former residence. The new house has a kitchen, two bedrooms, a hall, and a latrine. Raj's health has continued to improve, and he was able to get a good job in a nearby shop.

Father Joseph says that the family never dreamed of such a wonderful house. They are now happy and leading peaceful lives in the new home. The family feels that Jesus Christ helped them to get their new house.

Spirits Lifted, Hope Restored, Health Improves

In a country far from India—Honduras in Central America, former international partner Todd Zylstra-Garth tells a gripping story of a transformed family, with vastly improved health, brought about by their new Habitat house. Here is Todd's moving account of the matter:

I was an international partner in Honduras from 1989 until 1991. The latter part of that time was spent in San Isidro, in the Yure River area of northern Honduras. I got to know Doña Gloria who often did laundry for me and other IP's. Doña and her family lived in a shack of loose boards, about twelve by fourteen feet with a small leaky sleeping room attached. During harvest, their house was also the storage facility, and I can remember sitting on the corn—three feet deep over the entire floor—watching an old black-and-white television and shucking corn with the family.

Doña Gloria's family desperately wanted a Habitat house. Unfortunately, the Yure River Habitat affiliate covered many villages, and we were not working in San Isidro at that time.

When I left San Isidro in 1991, Doña Gloria's husband had hepatitis, the mattress on which he laid was wet from the leaky roof, the cow—their only real asset—had died and, of course, they lived in a miserable dirt-floor shack.

In 1993, I led a Global Village work team to San Isidro. One of the first things I did was go visit Doña Gloria and her family. What a change! Where the shack had stood was a new Habitat house! Doña beamed. Out front was an old pick-up truck. Somehow, her husband had managed to buy the vehicle and start a business of transporting families around the mountains for funerals, weddings, reunions, etc. His hepatitis had made work in the corn fields difficult so this truck had helped to generate more income for the family. And he was clearly much healthier now.

I went into the house. Everything was spotlessly clean. I was invited to sit in a comfortable, new-looking overstuffed chair. Doña could barely contain herself. I could barely contain myself. Doña Gloria's family was safe, dry, secure, and prospering. The children were especially bright and healthy.

This kind of story is being repeated thousands of times all around the world as more and more shacks are replaced with good, solid houses. Spirits are lifted, hope is restored, and health improves when families are able to leave leaky roofs, mud walls, and dirt floors behind and move into simple, decent places to live.

Chronic Asthma in a Moldy Trailer

And the story in North America is a different version of the same situation that is found in Africa, India, Central America, and elsewhere around the world. For example, consider the incredible story of the Browne family. The Brownes had a new house built for and with them by Penn-York Habitat for Humanity, headquartered in Sayre, Pennsylvania.

I first learned about this family when the mother, Amy, wrote a letter to Linda and me pouring out her heart in gratitude for their new Habitat house. She described their terrible housing situation before moving into the Habitat house. She said that they were living in a cold, drafty trailer with a leaky roof, full of mold and mildew. This deplorable place was very detrimental to their baby boy, Timmy, who suffered from chronic asthma. Amy said that she and her husband, Todd, had to give him medication and help him with breathing exercises every few hours just so he could survive. She said that they came close to losing him many times. Her heart was broken every day to see him so sick.

Then one day they saw Habitat for Humanity featured on the television program, *This Old House.* That prompted them to contact their local Habitat affiliate and apply for a house. Six months later the family was approved, and within that year they joyfully moved from the old trailer into their beautiful new Habitat house.

Amy said Timmy began to improve almost immediately. Within three months, she wrote, he was much healthier and more active. She said that because he was living in a clean, warm home, he became a strong little boy—not at all like the sickly child he had been.

Saved a Little Boy's Life

Amy concluded her emotional letter by stating boldly that she was sure the Habitat house had saved her son's life. For that reason she hoped one day to be able to meet us so that she could thank us for starting Habitat for Humanity and so she could hug our necks!

Somehow, a copy of her letter got to the regional office that covered Pennsylvania. In May of that year, Linda and I were scheduled to speak at the annual regional conference, which was to be held in Harrisburg. Unknown to us, arrangements were made for the Browne family to attend. At the big Habitation service at the conference, Robin Monaghan, administrative director of the region, read Amy's letter to the packed congregation. She broke down in tears a couple of times while sharing what Amy had written to us.

At the conclusion of the letter, Robin turned to us and said, "Millard and Linda, the Browne family is here tonight. I'm going to ask them to come up and give you that hug that Amy wrote about." Then she turned to the audience and called for the Brownes to come forward.

They were sitting in the back of the church. As the family—Amy, Todd, Timmy, Alex, and Amber—walked down the aisle, everybody started applauding. When they got down front, they walked up on the stage. And we started hugging! By that time, a dozen people were on their feet right in front of us, and flashbulbs were going off in rapid-fire fashion everywhere. The whole congregation rose to their feet and started cheering. It was an incredibly joyous scene.

Finally, all was quiet and everyone sat down. The family started back to their seats. As they were walking down the stairs from the stage, little Timmy exclaimed in a voice loud enough to be heard by nearly everyone, "I like this job!"

Linda and I have stayed in touch with this fine family. They continue to do well. Timmy's first two years of life were filled with hospital stays and breathing treatments. After moving into the new Habitat house, over a four-year period, he has not had any more asthma attacks.

A Sturdy Door

Amy says that their other children, Alex and Amber, are much happier and more secure in their new home. When they lived in the trailer, she said, Amber was afraid to sleep. She'd get scared by the terrible noise the rain made when it beat down on the trailer's tin roof. There were times when the wind would blow hard, the door would get blown open, and the roof seemed like it would just fly off. Amber was so frightened that she would cling to Amy and Todd.

When the family moved into their new house, the building coordinator knew that Amber was frightened by the rundown trailer, so he took Amber by the hand and gave her a personal private tour of the house. He showed her the locks on the windows and pointed out the sturdy new front door. He proved to her just how solid and safe the new home was. After that, Amber and Alex could sleep without worries.

Amy says that she and Todd have less worries now too. "Our children mean the world to us," she said, "and thanks to Habitat for Humanity we are now able to raise them right."

Best Thing—Our Children Are Well

A similar story to that of the Brownes comes from South Carolina. A mother of four whose family had recently moved into a new Habitat house spoke at the affiliate's Christmas party. She spoke with great conviction about the transforming effect of the house on the health of the children.

"I am very grateful for the beautiful home we have and the wonderful people who built it. But the best thing is that our children are well now," she said. "I thought I had very sickly children until I moved into my new home and found that their health improved remarkably in just a few months. Yes, they still have colds and the usual infections, but they are no longer chronically ill."

Yet another wonderful story of transformation and better health comes from Mount Pleasant, Texas. Roy and Penny Newton and their two daughters, Alexis and Kandis, were living in a mobile home out in the country. Penny graphically described the condition of their home: "The flooring was bad. The pipes were rusted, making the water bad. The place was cold. We did not have the necessary heating we needed to stay warm. And it rained inside the house."

At that time, said Penny, Kandis was quite ill, too sick to go to school and almost too weak to sit up and play or do anything. Penny was feeding her through a tube.

Then one day Penny saw something on television about Habitat for Humanity. She called the local affiliate, Mount Pleasant Habitat for Humanity, and after being accepted, a house was built with the family.

Penny reports that the house has been a great blessing to their family. She says that Kandis has improved tremendously and that the whole family is happier and healthier.

Environmentally Safe House

Quite an unusual Habitat house in Maine has had a dramatic effect on the health of a woman with a serious chemical sensitivity. Aroostook County Habitat for Humanity chose Leanna Moorehouse and her husband to have their first house. Building the house was a challenge because Leanna had had a work-related respiratory injury that had left her with chemical sensitivity. The house for her had to be "environmentally safe."

With the guidance of her doctor, the building committee selected materials for the inside of the house. Many of the sample materials from the manufacturers were sealed in plastic bags, which Leanna had to smell to test for sensitivity. Interior materials, such as plywood cabinets, were triple sealed with water-based urethane to prevent off-gassing. The vapor barrier that was used was double-faced aluminum foil blister-bubble pack with all seams sealed with a special aluminum tape. This was done to prevent any off-gassing from fiberglass insulation. An air-to-air heat-recovery exchanger was installed that completely changed the air in the house three and a half times every hour in the winter.

Leanna says that her respiratory attacks have been reduced greatly since being in the new house. She has not had to use her oxygen equipment in the more than two years she and her husband have been in the house. She has only had to take predisone therapy once and antibiotics twice. That was a dramatic reduction from her previous residence.

Leanna's respiratory injury is permanent, but living in her special Habitat house has greatly helped her and has given her much more control over her condition. Also she happily reports that the new house has benefited her husband in coping with his disabilities. "The house," she said, "has given us a secure environment in which to live."

Lead in His Blood

A healthy house in Springfield, Ohio, dramatically changed life for the better for the Sigmon family. Thelma and James had lived with their children for nineteen years in a rented house that had "floors ready to fall in and windows ready to fall out."

Six years earlier, a doctor discovered that their youngest son, Lester, had dangerously elevated levels of lead in his blood. Then only a toddler, he was already exhibiting classic behavioral symptoms of lead poisoning, probably caused by ingesting or inhaling lead-based paint that had peeled from the walls and ceilings of the old rented house.

The Sigmons urgently wanted to get out of the contaminated house. At that time, Thelma saw an article in their local newspaper about Habitat for Humanity. She attended a meeting and applied for a house. The affiliate—Clark County Community Habitat for Humanity—quickly got the house built. The family moved into their new home in October 1993.

Five years later, Lester is doing great. Out of the old house with lead-based paint and with continued treatments to lower his lead levels, he is an active little boy and a straight-A student.

House Was Killing Their Son

Rose and Kim Barnes know what a difference a good house makes. They believe their new Habitat house saved their son Dylan's life. In 1993, Rose and her husband, Kim, and their two children lived with his mother in Gerkish Township, Michigan. It was a very cramped situation, and his mother was suffering from emphysema. Plus Rose was expecting twins. Nicole and Dylan were born October 10, 1993, six weeks prematurely. Dylan was diagnosed with asthma. He was hospitalized numerous times for several months. In March 1994, he almost died. Rose said that he was allergic to almost everything: mold, dust, mildew, hair spray. Their doctor told Rose and Kim that the house was killing their son. The house was built right on the ground, and when it rained, the house smelled musty. The doctor told the Barnes that they needed to move out of the house or they would lose their son.

Rose and Kim learned about Habitat for Humanity from a local church pastor. They applied and were approved in January 1996. Several months later they moved into their new home.

Dylan's health began to improve immediately. In May 1998, Nicole and Dylan started kindergarten. When their pediatrician saw the youngsters for their kindergarten physical and shots, he exclaimed, "I can't believe this is the same Dylan. He's put on weight, and his eyes aren't black anymore."

Rose is totally convinced that their new house was the critical factor in Dylan's dramatic improvement. She exclaimed, "Since we moved into our Habitat house, Dylan hasn't set foot in a hospital."

Dora's Special Needs Foster Child

A touching story of how a new Habitat house *and* a lot of love made a huge difference in the life of a little boy comes from Gwinnett County Habitat for Humanity, just north of Atlanta.

Dora Bailey is a widow who raised four children on her own. As a missionary for her church, she went to a local children's hospital to pray for certain patients. She began praying for Marquis Hunter, a nine-year-old boy who was in a foster parent home after he was involved in a automobile accident that left him in a coma. Marquis was eventually transferred to a state hospital with no hope of recovery. That's when Dora Bailey became involved. After several visits, Marquis awoke from his coma and began responding to Dora's voice. Upon completion of foster-parent training for a special needs child, Dora took Marquis home with her. That "home" was a single-wide mobile home that leaked and needed repairs. Marquis's motorized wheelchair could not get through the door.

Dora applied for a Habitat house and was quickly accepted. Nurses and other staff from the children's hospital helped Dora complete her sweat-equity hours. In mid-1996, Dora and Marquis moved into their new Habitat home in Norcross, Georgia. The house is fully handicapped accessible. Marquis now attends school, and a computer has been provided for his use.

Dora praises God for bringing her and Marquis together. They both needed each other. She also praises God for giving them a Habitat house.

A Turnaround of Love and Special-Needs Health

Another touching story of compassionate people in a local Habitat affiliate helping a family through years of turmoil to better health and ultimate victory comes from Halifax Habitat for Humanity in Daytona Beach, Florida.

Albert Hill was injured on the job in 1992. He was rendered totally disabled. He was recovering from surgery that year when the family moved into their new Habitat house. Albert's wife, Carol had long been suffering with kidney problems. One of their five children, Samantha, was dropped at birth and had been without speech and bodily movement ever since.

Albert was receiving disability insurance for his family until 1994, when his insurance company went bankrupt. Habitat kept on top of the situation as the Hill family policy was shifted from agency to agency.

Dorothea Tonneson was the family partner, the Habitat member who worked with them. Dorothea, incidentally, is not only a fantastic Habitat partner in her local affiliate, but she also embraces and supports the worldwide work of Habitat for Humanity with great joy and enthusiasm.

Dorothea prayed with the family, loaned them money, and remained a faithful friend throughout their ordeal. Jane Walters, the executive director of Halifax Habitat, wrote to legislators, arranged legal counsel, and diligently sought community assistance. The Prince of Peace Catholic Church and community members paid many of the bills for the family until 1997. In that year, the Halifax board of directors set up a fund for the family. That fund then paid the family's bills until Albert finally was awarded compensation retroactively and started receiving regular biweekly checks for his family.

Carol's health has improved in the new house, and she started attending a local community college after being out of school for more than ten years.

Samantha, whose life expectancy was ten days at birth, is now a teenager. She is seen as a blessing by the entire Hill family, who work together to take care of and provide love for her. Although she has been bedridden, Samantha now has a special wheelchair. The county provides home educators for her. Her most recent, a music teacher, uses bed pads through which she makes music. Samantha responds positively to this music, which has given her a way to express herself.

By late 1997, the Hill family was financially stable and greatly enjoying each other and their Habitat house.

Literally Frozen inside Trailer Walls

A tornado ripped through a trailer park in Edmonton in the province of Alberta in western Canada and almost killed the Trendall family. Their trailer was demolished, but miraculously, no one in the family was injured. According to Brenda Trendall, "The hand of God was upon us."

After the tornado, Brenda, her husband, Doug, and their children moved to another trailer park. That new trailer had ice on the walls six months of the year. One winter morning the family was literally frozen within their house. They had to call neighbors to come and pry open the front door.

There was another compelling reason the Trendalls wanted to get out of the trailer. Their baby daughter, Chelsea, was born with Down syndrome and various other health problems. It was very difficult to practice the intervention techniques that she needed to help her develop basic skills in such a small, crowded space.

One day, Brenda and Doug heard about Habitat for Humanity on television. They contacted the Edmonton Habitat affiliate and applied for a house. The family was accepted. The house was built, and in 1997 the Trendalls—Brenda, Doug, and their children, Linda, Jaimie, Mindy, Dora, Cody, Rebecca, and Chelsea—moved from their crowded single-wide trailer with two bedrooms into a two-story house with room for everyone.

Chelsea is flourishing in the new house, as she has room to experiment. There is room to teach her how to crawl and go upstairs. She now makes intelligible sounds. Child-development workers are astonished at her progress.

Other members of the family are also doing well. Everyone is much more relaxed in their spacious Habitat house. As with most families, there is less arguing now that there is room to have private spaces, and the children are doing well in school since they haven't missed any days because of sickness.

These are only a few of the thousands of stories about the healthy qualities of a decent place to live. We take for granted the "perks" that come with our good homes until we hear such stories. And what more essential perk do we need to live decently than good health? Good houses and good health. They do go together, and it's a wonderful thing to know as we house-build. We never forget this essential "more than houses" element as we build it in with each solid wall and reliable roof.

CHAPTER 7

Launching New Careers

I didn't just receive a clean, healthy, and beautiful home;
I received a new me!
—Teresa Burton, Habitat homeowner

Every aspect of the work of Habitat for Humanity is about change. People change. Lives improve. Children do better in school. Health, both physical and psychological, grows stronger. Neighborhoods blossom. New relationships emerge and existing ones are strengthened.

But one dimension of the change that takes place in this hammer-and-nail ministry is that of launching people—homeowners and others involved—into new careers or activities that are so much better for them and for their respective communities.

Consider, for example, the case of Karlee and Gary Bohn of Anchorage, Alaska. Linda and I met this fine young couple and their six boys when we were visiting Habitat houses in that state in August 1998. We heard Karlee speak at a luncheon for pastors and other church leaders and then visited in their home later in the week.

New House, New Job, New Future

Karlee and Gary were evicted from their apartment without warning, so they became nearly homeless, living in a tent part time then in a two-bedroom apartment. Gary was sporadically employed by a company that frequently took him out of town to a remote site, leaving Karlee at home struggling to keep the boys quiet to avoid hassles with neighbors.

Then, in early 1996, the Bohn family was selected to receive a new

Habitat for Humanity house. Their sponsor was a group of Lutheran churches in Anchorage. A joyous groundbreaking was held on May 21.

Karlee said that it was amazing for her family to be a part of building their dream home from the ground up. The whole family participated in putting in their five hundred sweat-equity hours, but Gary was the major contributor. He was on-site throughout the whole process.

The family moved into their new home in October 1996. On that occasion, their nine-year-old son, Jess, exclaimed, "We're just like rich people. We've got a house and everything!"

Karlee spoke movingly at the luncheon in August about the very good things that have happened to her family since they moved into their new home. "Because of Gary's experience in building our house, he was chosen over three other people to get a good job here in town at a lumber company that sells everything to build a house. The owner was impressed with the fact that my husband had just gone through all the steps in building his own house. Now he is in charge of the whole lumberyard."

Karlee continued, her voice cracking with emotion, "Last summer, we received a beautification certificate from the Mount View Community Council for our property. It's a great feeling to take pride in a house and yard and have them reflect out into the neighborhood and community.

"Having a home and my family settled has also given me the opportunity to go back to college. I'm happy to say I'm now in my second quarter."

The children have also blossomed. One son has completed high school and is in training with the Job Corps. The others are attending high school, middle school, and grade school. A long-deferred dream of the boys has also been realized—since they now have a backyard with a strong chain-link fence around it and a dog running around back there.

In Arizona, a volunteer benefited from learning building skills working with Valley of the Sun Habitat for Humanity in Phoenix. Lucinda Smith said that she had become "somewhat" skilled at roofing, carpentry, and dry-walling. "Recently," she wrote, "I traveled to San Luis, Sonora, Mexico, with a church group. The church supports an orphanage which is home to fifty children. Although I was asked to serve as an interpreter, I worked with the construction team. We set about replacing wall receptacles, repairing ceiling fans, sealing doors and windows. My Habitat skills were invaluable. I was able to pass along knowledge to my teenage helper, who had never lifted a hammer. It was a joy to bring this gift of building skills into another country," Lucinda concluded, "and also a blessing to impart skills to others."

New Career, Second Career, Specialized Career

Meanwhile, back on the East Coast, in Gettysburg, Pennsylvania, a skilled builder named Carl Bailey was volunteering on a blitz-build in early 1997 with the local affiliate, Adams County Habitat for Humanity. One day during the build, a fellow volunteer asked Carl what he did for a living. He replied that he was self-employed as a contractor but that he really wasn't happy in that work. *Maybe,* he mused, *God has something better for me.* What he really loved doing was building with Habitat, he admitted to a fellow volunteer.

"Well, there is an opening in Harrisburg," the volunteer said.

Carl applied and was selected as the construction manager of Habitat for Humanity of the Greater Harrisburg Area. Carl is convinced God arranged the job for him.

For Dennis Burnett, Habitat for Humanity became a blessing and an avenue to a new career just when he thought his productive years of work were over.

Dennis worked for a small construction firm in Santa Rosa, California. Then his wife's job was transferred to Bloomington, Illinois. There Dennis found employment with a small contractor.

Six months later, Dennis developed intercranial hypertension and began to lose his eyesight. Within six weeks, he was legally blind. Although he could still see to function, he could neither drive nor read intricate documents, such as blueprints.

Out of work, Dennis felt depressed. Life as he had known it was over.

Then he heard about Habitat for Humanity. He phoned the office of Habitat for Humanity of McLean County and was given the name of Carroll Oien, the project director for a house currently under construction.

Dennis's son, Rob, drove him to the construction site, where he was put to work immediately. Dennis was overjoyed to find an outlet for his many construction skills.

After completion of that house, Dennis continued to work on the home maintenance team, helping to put the finishing touches on Habitat houses during the week. On Saturdays, he worked on other new Habitat houses.

In the three years since Dennis made his first phone call to the Habitat office, he has been project director or codirector on three Habitat houses and has provided technical assistance on many others. "Habitat for Humanity has been a blessing for me," he said. "It helped me get back on my feet emotionally and physically. I have received as much benefit as the

people for whom I've built houses." Judy Stone of the McLean County affiliate said that Dennis is one of the most popular directors with the Habitat
families in Bloomington.

Kelle Shultz, executive director of Knoxville, Tennessee, Habitat for
Humanity had a life-changing experience on a Global Village work team to
Nicaragua in 1993, which led to her new career with Habitat for Humanity.
This is what she shared:

> Life there is so different. It took me several days to realize the vil-
> lagers were happy, even though their days consisted of sleeping, get-
> ting up, existing, and sleeping again. They had no errands to run, no
> malls to escape to, no dry cleaning to pick up, not even a get-together
> with friends for dinner and a movie. I can't imagine working hard all
> day for eighty cents and then not being paid for over three months.
> Like Guiermo, one of the homeowners, said, these houses are their
> life, even though they are struggling to pay the six dollars a month for
> house payments.
>
> Imagine not being able to send your children to school or provide
> food other than the rice given to you by the local aid agencies or the
> bananas on the banana plants. The memory of Candida, another
> homeowner, asking Mike to count the "cucharos" at the end of the
> day because she could not count to seven remains vivid. Or, every
> morning seeing the local women carrying forty-gallon water buckets
> on their heads to the work-site and then work by our side without
> work gloves or shoes but always with a smile.
>
> My most poignant memory is our last night there. We had a dance
> to celebrate the conclusion of our work. After several salsas, the music
> and mood changed to a slow dance, and I found myself with six chil-
> dren wrapped around my legs, their arms entwined around me and
> themselves. Our whole Habitat group laughed as we looked around
> at each other. We all had at least four children hanging on at least one
> limb. I had never felt so popular. I looked down to see six sets of
> brown eyes and six smiles looking up at me with complete trust and
> adoration. I started to cry, and the children hugged me tighter. They
> were trying to comfort me!
>
> The people of Nicaragua made me see the importance of life. I was
> given the kind of love from these people that Christ gave—human to
> human, uncaring of faults or attitudes, just working hands and car-
> ing hearts. I left on my trip planning on helping folks build houses,

and they ended up giving me so much more; they helped me build my life.

Kelle explained that the trip changed her life, making her realize the importance of the ministry of Habitat for Humanity. A year later, she was hired as the executive director of Knoxville Habitat for Humanity. She continues to serve in that position and has been one of the driving forces in making Knoxville Habitat one of the great Habitat affiliates in the country.

Life-Changing Goal

Another person who had a life-changing experience with Habitat for Humanity is Karin Weaver. She was a student at Rhodes College in Memphis, Tennessee, in 1988 when she first learned of Habitat for Humanity. A friend suggested that she apply at the Memphis Habitat affiliate for the volunteer coordinator position. She did and was hired. Karin said that the year she spent at that job changed her life. She switched her career from potential attorney to low-income housing advocate.

Since that time, Karin has worked as a nonprofit housing program director, a mortgage lender for low-income borrowers, a Habitat committee member, and since October 1995 as associate regional director for the South Central Region (Kentucky and Tennessee) of Habitat for Humanity International. Karin received her real estate license, earned a certificate in nonprofit management, and is contemplating going for a master's degree in the same field.

"By working with boards, committees, staff, and volunteers," she said, "I feel I've been able to help the folks on the front line do a better job of meeting community needs. Throughout the past ten years, I've seen a special type of excitement develop in families who thought they'd never be able to own a house then realize that they are finally able to buy a home of their own.

"I was recently asked what I want to be proudest of when I reach my eightieth birthday," she added. "I replied that I want to look back and know that I helped hundreds of families achieve the dream of homeownership. Habitat is allowing me to walk that path, and I am grateful. Thank you for helping me reach my goals by helping others reach theirs."

All across the country and around the world, dedicated people like Karin certainly are helping families reach their goals and realize their dreams. And this is true not only in terms of better housing but in the fulfillment of so many other goals and dreams.

Habitat Owners and Better Jobs

The AREA study mentioned earlier stated that 40 percent of the homeowners interviewed had changed jobs since moving into their new Habitat homes. Twenty-five percent of those claimed that the opportunities available at the new job were greater than those at their old job.

"In general," the study reported, "homeowners who claimed that homeownership has positively affected their employment situation attributed their success in part to the reliability of constant house payments, which they did not anticipate changing over time. Some had returned to school or taken the time to learn a new trade, which had positively affected their future earning potential or their employment outlook. Others had been able to take off time needed to find a job that offered advantages over the previous job. One homeowner said that skills learned through building his house and other Habitat houses enabled him to become an apprentice woodworker. Habitat gave him the skill to start a new and better career."

Rosie Simmons of Chicago, mentioned earlier, is a dynamic example of a Habitat homeowner who has blossomed in a new career since she moved into her new house. Rosie was in a minimum-wage job when she and her four children moved into their Habitat house in north Chicago in 1989.

Soon thereafter, Rosie got an entry-level job at a bank in downtown Chicago. Over the next two years, Rosie was promoted four times. Rosie continues to work at the bank as customer service specialist in the ATM department. She also has become a major spokesperson for Habitat for Humanity in the metro Chicago area, giving frequent speeches about Habitat's work at churches, civic clubs, and other groups.

Henry Robinson got a job of a different kind after he and his family moved into their Habitat house on John's Island, South Carolina, in 1982. He, his wife, Viola, and their children, Davon and Michelle, were living in a trailer until it burned to the ground. Happily, no one was hurt, but they had no place to live. Henry and Viola applied to Sea Island Habitat for Humanity and were accepted to be a partner family with that affiliate.

Henry's occupation as a dry-wall finisher allowed him to do all of the dry-wall work on their Habitat house as a part of the sweat-equity requirement. Afterward, he was hired to do the dry-wall work on other Habitat houses. In the ensuing years, Henry has done the plasterwork on more than fifty Habitat houses built by Sea Island Habitat for Humanity.

Henry also continues to volunteer to do other work on Habitat houses. In fact, he has helped on nearly every Habitat house completed by the local affiliate since 1982. He paints, puts on shingles, trains other volunteers, digs footers—in short, he does whatever is needed to get the job done.

A Teacher's Extra "Career"

Not only in the United States are people finding new work and new meaning in their lives after moving into a Habitat house. Steve Weir, our Asia and South Pacific director of Habitat for Humanity International, and Murph Eliot, past international partner in Sri Lanka, shared the exciting story of transformation and generosity of S. Duraifaj. Through Habitat, he launched a new career of helping others after moving into his new Habitat house. Duraifaj was a teacher in central Sri Lanka in the town of Hatton. He earned about seventy-five dollars a month.

Duraifaj grew up in a "line house" on a tea plantation. Line houses are dwellings provided by the plantation owners, where families, often a dozen people from three generations, live in a single room. These rooms are attached to one another in a long line with no outside light or ventilation other than the entry doors lined up across the front of the building resembling cattle stalls. Often ten to fifteen families live in one of these line houses, and they share a single outdoor toilet and a single water spigot.

Fortunately, the Durairaj family was able to obtain an eight-hundred-square-foot plot of land. Durairaj wanted to build a tiny house for his family on that land, but he was repeatedly turned down by banks and other groups. Eventually, he heard about Habitat for Humanity, applied for a Habitat house, and was accepted. The groundbreaking ceremony took place on April 24, 1995. The family moved into a small temporary hut on their new property so they could make three thousand clay blocks, their sweat-equity requirement for the new house. The dedication service for the completed Habitat house took place on June 10, 1995.

A private landowner who attended the house dedication service was very impressed with Habitat for Humanity's mission and the changes in the Durairaj family. Subsequently, he offered to sell two acres of his own land at a very reasonable price so more needy families could have their own homes.

Durairaj formed an organization to buy the land. It was called Anbu Nagar, which means "Village of Love." Twenty people joined the new group, and they pooled their money to purchase the two-acre plot. Once purchased, the land was divided into twenty plots, with a small section set aside for Durairaj as appreciation for his help in making the deal possible.

Durairaj sold the parcel he had been given and received enough money to pay off his mortgage. But that is not what he and his family decided to do. Instead, they donated the money to the local Habitat for Humanity affiliate that had built their house. The Durairaj family wanted another family to receive the same blessing they had received.

Durairaj now serves on the family selection committee for Hatton Habitat for Humanity. When asked what his future contributions to Habitat for Humanity would be, he replied, "My ambition is to help others through Habitat for Humanity. I cannot do enough for this ministry."

One example of Durairaj's continuing help to others is a widow with eleven children, known only as Mrs. Susilawathie. She works as a street sweeper, making about fifty dollars a month. With Durairaj's help, she was able to buy a plot of land and apply for a Habitat house. After she was accepted, the blocks for the new houses needed to be made, but the block-making machine could not be delivered to the new site, which is situated on difficult terrain. So the blocks had to be made at the old house. Once the job was completed, Durairaj organized a group of teachers and students at his school to carry the blocks down the hill to the site for the new house. Six teachers and forty-three students made up this "block brigade."

When Linda and I visited Hatton in July 1998, we met Durairaj. He is a very shy and unassuming man but full of love and a spirit of giving back.

Space to Launch a Business

While in Hatton, we spent the night with another family that had recently moved into their new Habitat house. The father of that family, Bernard Gabriel, made his living by repairing motorbikes, which are quite common in Sri Lanka. He was so excited about his new Habitat house because it was such a vast improvement over the shack where he and his family had been living. But he was doubly excited because the lot was large enough to enable him to build a small garage for his motorbike repair business.

Other Habitat families have been able to launch businesses or expand an enterprise as a result of getting a Habitat house. Mbomba, my work foreman when Linda and I were living in Zaire in the seventies, started a small furniture-making operation in the backyard of his new house, using scraps from the lumber of other houses under construction. That business eventually closed, but his experience in running it gave him the confidence he needed to launch a much bigger farming operation that twenty years later was still going strong.

Not Just for Homeowners

Habitat for Humanity has also been the catalyst for launching new jobs or careers for literally thousands of people who are not homeowners. Following are the stories of a few of those people.

George Kampango of Mangochi, Malawi, in southern Africa, said, "Working with Habitat for Humanity has led me to gain a new career because I started as a store clerk and currently am working as an affiliate manager."

Catherine Ndunda, a twenty-five-year-old single mother, living in Maua, Kenya, east Africa, has been a part of Habitat for Humanity since 1995. She was trained as a project manager and assigned to work with Athiru Gaiti Habitat for Humanity. Catherine stated, "What a joy to do the work that suits you, offering the best services to the right people, and the fruits of my labor are blessed by Almighty God, the homeowners, community members, and my immediate supervisors. Facing challenges in fieldwork has widely increased my wisdom, and I always have an answer at hand to questions concerning Habitat work and other related ministries. I have really gained courage," she explained. "Interacting with all types of people in Habitat work has improved my personality, and I express my ideas without fear when need be, both to individuals and to multitudes of people. Working in different environments has made me adaptable to difficult conditions. I have become physically fit because of climbing mountains and going down valleys from family selection visits and house construction work. I have learned never to despair because I have faith and hope."

"I have been able to identify my talents, one by one," Catherine continued. "I am proud to work with Habitat for Humanity for whom I make use of my talents as well as practicing my hobby of socializing. I have gathered technical skills at construction sites. Although I am a woman, I can put up a simple, decent, and affordable house using locally available resources, so there will never be a time I will sleep outside because of a lack of shelter. I will always offer the best I can, physically, mentally, socially, and financially. Long live Habitat!"

"Their Love Cured Me"

Another incredible story of transformation following involvement with Habitat for Humanity comes from Canada. Brad Henderson of Ottawa, Ontario, shared his story with us.

"In February 1994, my wife, Debbie, and I were busy with our life, friends, and work. Debbie was working for the Canadian government, and I was a partner in one of Ottawa's largest insurance brokerage firms. I had started with the firm six years earlier when I convinced them that they should hire me to work as a commission salesman. During those years I built a significant portfolio of clients, and through hard work, I became one of six partners," he said.

"However, during a monthly partnership meeting, I had a strong difference of opinion with the senior partner–president. That evening, while talking to Debbie, I realized that I had to leave the business. What happened after that was a downward emotional spiral that left me, after four months, telling Debbie that 'I don't think that I will ever find work again that I can get excited about!' I realized that who I 'was' and what I 'was' was tied very closely to my work and my status in the community. When this was taken away, I felt empty and lost. The Bible warns us of belonging to the world, and I was an example of it," he said.

"What was I going to do? I tried going to a therapist, but it felt self-indulgent and foolish talking about my concerns to a stranger. I prayed for guidance and strength but nothing happened. I could not talk to my fellow business partners as I quickly became emotional and angry. I had always been enthusiastic and energetic, but my purpose and drive were gone. I was an empty balloon," he admitted.

"During that spring the local affiliate of Habitat for Humanity was preparing for its initial build, a two-family home. Debbie and I had become interested in Habitat while reading *No More Shacks*. After hearing a presentation at our church about the planned build, we signed up. We helped prepare for the construction. Then, when the time came to build, we took a week off from work to participate in the construction. Monday arrived with a beautiful sunny morning, and after the devotion we picked up our hammers. I cannot quite describe what happened that week while we worked. With each nail I hammered, I was being refueled. My anger and frustration were being replaced with hope and confidence. The people I met were an inspiration, giving of their time to help others. Their love cured me, and I am grateful to them and Habitat for providing this community where I could heal," he said.

"The following Monday I returned to work and immediately went to the president and told him calmly, without anger, that we had to resolve a number of matters as I was determined to leave the firm. In the following month we answered a job offer to work in rural Ecuador for a Canadian development organization," he went on. "God, knowing that I was healed and ready, opened a number of doors that we walked through, and by the following January, we arrived in South America to start our work there. After finishing our three years' commitment in Ecuador, we approached Habitat for Humanity International in 1998 and were selected to return to Latin America to work in Bolivia."

Brad and Debbie are serving in the Santa Cruz Habitat affiliate in Bolivia. The Hendersons say that they feel blessed to be a part of the ministry of Habitat for Humanity. Habitat certainly is blessed to have them.

An Avid Avocation

In Michigan, a United Methodist pastor, Ken Bensen, has been profoundly affected by Habitat for Humanity, and he has made an awesome impact on Habitat. I met Ken in 1985, when he and his wife, Bernice, were living and working in Chicago. Ken was director of Metro Chicago Habitat for Humanity. He served in that capacity for two years and then moved back to his native Michigan to become pastor of Faith United Methodist Church in Lansing.

Ken quickly involved himself in the work of Habitat in Michigan. In

The Reverend Ken Bensen speaks at a house dedication in Montmorency, Michigan.

1992, he formed a statewide Habitat organization and was named its president. Then, methodically, he proceeded to plant local Habitat affiliates all over the state. When Ken arrived in Michigan, there were six Habitat affiliates. By early 1999, there were eighty-four, making Michigan the state with the most Habitat affiliates in the United States.

Ken is constantly organizing various events to promote and expand the work. In 1998, thirty Michigan affiliates participated in Building on Faith week, the most affiliates of any state. In the spring of 1999, Michigan affiliates sponsored and built four houses in the Jimmy Carter blitz-build in the Philippines. Ken and eighty-five other people from Michigan went to the Philippines to build. In June 1999 a one hundred and eighty house statewide blitz was held. That ambitious build was called the Jack Kemp Blitz-Build. Jack visited several of the sites and helped with the building.

Ken has also arranged for a prison in Saginaw to build several Habitat houses (and has plans for more prisons to also build), and he is organizing "Re-Stores" in affiliates statewide to receive and resell construction materials as a way to generate more money for building Habitat houses.

All of this is being done while Ken continues to service his three-hundred-member congregation. But this second "career" of Ken Bensen is so deeply meaningful to him. He has told me that he feels that he truly is doing the work of Christ through Habitat for Humanity. He certainly is blessing thousands of people through his effective work even as he is being blessed.

Habitat Careers

Sometimes people come to Habitat as volunteers and stay until Habitat becomes their careers. Elmer Farlow of Sophia, North Carolina, is such a person. He has been heavily affected by Habitat for Humanity and, like Ken Bensen, has left a deep mark on this ministry. In Elmer's case, his Habitat avocation became his nearly all-consuming vocation.

Elmer had his first involvement with Habitat for Humanity while at Mars Hill College near Asheville, North Carolina. The local Habitat affiliate, Madison County Habitat for Humanity, was just getting underway. Elmer started going out on Saturdays to help with the building of their first house. In that time, he learned about Habitat's mission and Christian roots. Elmer felt that the work fell in with God's call for his life.

After graduating from college in 1990, Elmer decided to contact Habitat International about volunteer opportunities. He visited Habitat headquarters in Americus in January 1991. While there, he learned of an opportunity to be a construction superintendent with the annual collegiate challenge program. Collegiate challenge links up thousands of college students with local affiliates so they can use their spring breaks building Habitat houses. Elmer spent three months in Sumter, South Carolina, coordinating the work of several hundred students from fifteen colleges and universities who came to build with Sumter Habitat for Humanity. Returning to Americus, Elmer was informed about the fifteenth anniversary celebration of Habitat for Humanity, which would be held in Columbus, Ohio, in the fall. Fifteen work teams were to start from various points throughout the United States and build for fifteen weeks with fifteen different affiliates as they all moved steadily toward Columbus for the celebration—building along the way.

Elmer was accepted as one of the team leaders for the nationwide moving blitz-build. His team started in Omaha, Nebraska, and worked its way through Nebraska, Kansas, Missouri, Illinois, Indiana, Kentucky, and Ohio. A fifteen-house blitz-build was scheduled in Lexington, Kentucky. Actor Paul Newman joined in to work on that exciting build.

After the Columbus celebration, Elmer worked for a while as a part-time construction purchasing agent with the Greensboro, North Carolina, affiliate. Then, in April 1992, Elmer launched a two-year stint as the construction superintendent for the High Point affiliate. He helped that dynamic group launch a seventy-house subdivision.

Even as he worked full time in High Point, he continued to travel to other sites to work with Habitat. In June 1992, Elmer participated in the Washington, D.C., Jimmy Carter work project of ten houses. In 1993, he came back to Americus to participate in our twenty–twenty thousand build where twenty houses were built in a week, including the milestone twenty thousandth house. In 1994, Elmer resigned his staff position with High Point Habitat for Humanity but was elected to serve on their board of directors.

In addition to his extensive work with Habitat, Elmer has served in many locations in the United States and abroad for his church and other Christian organizations. On one such trip to Rwanda in 1994, with a medical team, he met a nurse named Barbara. They were married on March 25, 1995.

Barbara has the same heart as Elmer for service in the name of Christ. The two of them have gone to Antigua, Honduras, and to several locations in North Carolina in connection with disaster relief. And Elmer continues to serve Habitat for Humanity in numerous ways. Elmer said of his service career, "It has been quite a ride with many people helping along the way through prayer and financial support. The ride has not been without a few rough times, but Jesus Christ has been there to lead me all the way. I look forward with anticipation to what God may have for Barbara and me to do in the years to come."

The Energizer Bunny

Another person who has had a spectacular Habitat career is Randle Dew. The newspaper in Arcadia, Florida, calls him "the energizer bunny."

When Randle retired from the pastoral ministry in 1982, he moved to Paducah, Kentucky, to be close to his grandchildren. After settling in, he joined up with Paducah Habitat for Humanity, one of the very first Habitat affiliates in the country. The group had just completed their first house, but things were not going well.

At the first meeting he attended, the president resigned, and the only other board member at the meeting offered Randle the job. Randle declined but offered to become a non-paid executive director. For the next eight years, Randle worked as a volunteer with Paducah Habitat for Humanity. During that time the affiliate built eighteen more houses. Randle said that his experiences with Habitat changed his life.

He planned the first blitz-build of a Habitat house in Kentucky. He also set up the "Habitat Hilton" at a local church that provided a place to sleep and an indoor pool for work

Women inmates get on-the-job training by constructing Habitat houses in Shelby County, Tennessee, as part of their Women in Construction Building for the Future program. Photo by Dr. Ernest Withers.

campers who came to Paducah to build Habitat houses. Randle helped start Habitat affiliates in the surrounding counties, too.

When Randle *really* retired in 1993 and moved to a mobile home park in southern Florida, what did he do? He started another Habitat affiliate in Arcadia. That's why the paper there called him "the energizer bunny." Randle is full of energy, and he uses that energy to the great benefit of Habitat. We love him for it.

"This Is More Than a House"—One College's Experience

Yet another man who has blessed Habitat for Humanity and who has seen both Habitat homeowners and student volunteers transformed is Wayne Nickerson, dean of the chapel at Westminster College in New Wilmington, Delaware. Wayne has known many students whose lives were never the same after their experiences with Habitat house-building.

"The receiving of a decent home and the process of participating in the construction of your new home—both experiences change lives," Wayne said. Each building project has an impact on everyone involved. As he put it, speaking for each house built, "This is more than a house."

Sometimes the volunteers have their lives changed by those with and for whom they are building a house. In January 1992, Westminster students were among the first to work on Habitat houses in Alta Verpaz, in north central Guatemala. Working on the very first house to be built in the city of Coban was a student named Matthew Barnishine.

Matthew was approached one day by a local man, a deacon in the Nazarene Church. The deacon took a small cross from his neck and placed it into Matthew's hands. Matthew handed the cross back and said he could not wear it. The man simply smiled and closed the young man's hands around the cross and said, "Just keep it. I will pray that the day will come that you can wear this cross and have it mean something to you."

That moment was a turning point for Matthew. Today he wears the cross as he teaches at a Christian high school in western Ohio.

Kevin Knab was also on that 1992 trip to Guatemala. Recently ordained as a Presbyterian minister, he traces his awareness of human need, particularly in Central America, to his experience with Habitat for Humanity.

"Christ calls us to love, which is seemingly an easy task," wrote Kevin. "I think the most powerful impact that Habitat had on me is that it taught me the depth and commitment that are really involved in genuine love. This began to occur through a developing awareness of the world around me that I had never had before. Besides the grace that one receives in working beside those who will move into their new homes, which is such a part of everybody's Habitat experience, I also had an awakening that spurred me to question the circumstances that put the working poor in this position in the first place. My experience with Habitat taught me that love of those in despair is not something you can choose to turn on during short mission trips and turn off when you return home. It involves serious love, which involves serious life commitments. Otherwise the work is in vain. The experiences I had with Habitat in high school and college are with me now as a Presbyterian minister heading back to Guatemala with Habitat for Humanity this very month."

Wayne Nickerson shared the transforming effect on three other students, men who found their calling in life after Habitat trips. "Dan Jones, on a 1995 visit to Guatemala, is confirmed in his calling as a doctor," he explained, "a doctor who wants to bring healing to the disposed of the world." Keith Bittel, also a 1995 volunteer at Alta Verpaz, found his calling to teach in an urban high school. And Scott Mulrooney, as a result of his experiences with Habitat while a Westminster student, came to Americus to be trained as an international partner. He then went to Guatemala and served two years then moved to Antigua to launch the first Habitat affiliate in that Caribbean island nation. In July 1998, he joined the staff of our Washington, D.C., Habitat International office to handle international matters.

On summing up his work with Westminster students and Habitat, Wayne Nickerson made this powerful statement: "I can only say that as a person who has worked with college-aged people for twenty-five years, Habitat for Humanity has been the single most influential experience in my attempts to change student lives."

From Japan to Habitat to Sea Island

Many college students are surprised at what they find when they build a Habitat home. Norihiro Yamazaki is a young man from Japan. He came to the United States in 1995 to study at Syracuse University. Soon after arriving at the university, he became involved in the campus chapter of Habitat for Humanity. Every weekend he worked with the local Habitat affiliate. He also participated in two spring-break trips—one to Mississippi and another to Kansas.

Through that work, Norihiro decided that Habitat for Humanity was "a little different from other volunteer organizations. The purpose and the method are well connected in this organization," he said. "The goal is to help low-income families, and the method is to provide houses to them at affordable prices. I helped to build a lot of houses. After we built them, homeowners thanked us. When I saw the faces of these new owners, I felt great satisfaction. I felt that what I did helped people and made them happy. This fact makes me think that Habitat is a great volunteer organization."

Because of his wonderful experiences with Habitat for Humanity, Norihiro decided to volunteer for a year with Habitat before launching his business career. He was accepted by Sea Island Habitat for Humanity and began to serve at John's Island, South Carolina, in late 1997 in a stipend construction position.

Surprising Second Careers

A husband-and-wife team in Norfolk, Virginia, have found a deeply rewarding second career in building Habitat houses with South Hampton Roads Habitat for Humanity. Sylvia Hallock, executive director of that affiliate, believes the contribution of Dan and Terry Anglim is unique. Dan had a successful career in the navy and retired with the rank of captain.

In 1992, the Anglims saw a newspaper advertisement about a Habitat for Humanity meeting. They went to that meeting and got hooked on Habitat. Every year since, they have built one or two Habitat houses. At the beginning, they knew nothing about building a house. Over the years, they have become self-taught building experts. Dan said that he just read construction books, always staying one chapter ahead of where they were in the actual building process.

The Anglims not only build the houses, however. They also solicit building materials and raise funds. Dan and Terry view their work as an application of the "theology of the hammer—God's work, our hands."

As for Edgar Stoesz, he was surprised by the turn his career took toward Habitat. Edgar's story begins on a farm during the Depression. His teen years were the years of World War II, followed by the Korean conflict. True to his Mennonite teachings, he registered as a conscientious objector. Newly married to Gladys and with a child on the way, Edgar entered service with the Mennonite Central Committee to do two years of alternative service.

His intent was to get this obligation behind him and then get on with the rest of his life. But he continued with the Mennonite Central Committee for thirty-two more years beyond the required two. Edgar's responsibilities with MCC broadened. He was stimulated and felt fulfilled by his work. Then a reorganization was done at MCC, and Edgar found himself without a job. He was devastated. The situation seemed utterly unfair and made no sense. In his midfifties, he was confident that he still had much to offer. Naturally, Edgar was confused and angry. He realized he was slipping into a medium depression. On a dreary winter day, sitting in his gloomy basement office, he took the back side of an envelope and, with a crayon, wrote, "Do you have the faith to believe that the best is yet to be?" Edgar thought it was hardly imaginable and didn't know why he wrote that. The past had been so good—the future seemed so bleak. But Edgar put the envelope in a prominent place, hoping it might inspire him on another occasion. Later, still feeling miserable, he angrily crossed out *faith* and replaced it with *audacity*.

But something soon happened. All alone, driving somewhere in Virginia, he started to pray, "Lord, open the way." And suddenly his mind cleared. He envisioned a bright future with a focus on family, church, and the poor. The poor? How could he best serve them? Edgar suddenly thought of Habitat.

Edgar and I had met several years earlier, and I had questioned him about going on our board. At that time, he did not feel that his busy schedule would permit such service. Now Edgar was available. He called fellow Mennonite LeRoy Troyer, who had been actively involved with Habitat for Humanity and was serving on the international board. Soon LeRoy talked to me about Edgar and about Edgar's desire to serve with Habitat for Humanity. I asked Warren Sawyer, chair of our international program committee, to invite Edgar to serve on that committee as a way to get him involved. Edgar accepted.

He said that the relationship seemed like a fit from the outset. Edgar was able repeatedly to bring his former experiences to bear on Habitat's growing international program. Life was beginning to make sense again. Two years later, Edgar was elected to the Habitat international board, and two years after that he was elected to be chairman of the board. Edgar served for a total of seven years on the board, four as chairman, and he served incredibly well. During his tenure, the organization grew exponentially, and Edgar helped guide the growth faithfully and expertly.

Here's what Edgar had to say about his experience with Habitat: "My life was changed by my service on the Habitat International board, and I will always be grateful to God and the wonderful board and staff persons with whom I was privileged to serve. It was, as the Scriptures say, in losing my life in service to others that I found it in a richer and deeper sense."

New Start in Los Angeles

Like Edgar, Karla Alltizer found an exciting future with Habitat for Humanity after a devastating personal crisis. She was in the midst of counseling sessions trying to recover from a painful divorce following twenty-six years of marriage. Her counselor advised Karla to do some volunteer work as a way to get her focus off her pain and onto something positive.

Karla phoned Los Angeles Habitat and learned that they were just getting ready to begin building their first house. She started going out to the building site on Bliss Street in the Watts-Willowbrook area. Karla said that she quickly got caught up in the excitement of the Muniz family as their dream of homeownership was becoming a reality. "It was just the beginning of the great experience of Habitat," she said. "I was hooked!"

Karla was faithful and diligent in her work with Los Angeles Habitat. She helped build several houses and made lasting friendships. "A healing was taking place for me," she wrote, "so I reached for more by joining the family selection committee. Now I was not only helping to build the houses but also helping to choose the families that would occupy them. It truly was a spiritual experience and showed me the meaning of the Bible scripture, 'a person is justified by his works, not by faith alone.'"

In June 1995, Jimmy Carter held his annual blitz-build in Los Angeles. Twenty-one houses were to be built in a week in Watts. Karla took vacation time from her full-time job to fully participate in that exciting week of building. Four of "her" families were among the twenty-one homeowners. She was thrilled to work alongside them and the hundreds of volunteers who came to make a difference that week. The dedication services at the end of the week, with the presentation of the Bibles, was the emotional highlight for Karla. She wiped tears from her eyes along with other volunteers and the homeowners as they received their Bibles and keys to new homes and new lives.

After the blitz-build in Watts, Karla decided to switch to the San Fernando–Santa Clarita Valleys Habitat affiliate since it was closer to her house. She became chair of the family selection committee. Soon thereafter, Karla had an unexpected turn of receiving from Habitat after years of giving. Her oldest daughter, Kristin, had moved to Oregon and had been abandoned by her husband. She was left with four children. A few years later,

Kristin met and married Chuck. Two more children were born. The eight of them were living in a two-and-a-half-bedroom house. And their income was not sufficient to get conventional financing to purchase an adequate home.

The family applied to Coos Bay, Oregon, Habitat for Humanity and was selected to have a Habitat home. On March 8, 1998, the completed home was dedicated, and the family moved in. Karla was there, completing a circle of giving and receiving. And you can be sure the tears were flowing again—tears of happiness and great joy.

A Roof for One Woman—A New Future for Another

In 1991, Olive Jean Bailey, a member of Plymouth Congregational Church in Minneapolis, Minnesota, went on a work camp to Coahoma, Mississippi, to help build Habitat houses. Upon her return from Mississippi, Olive Jean shared her enthusiasm about the work in Mississippi with Plymouth Church. Holly Vincent was impressed with what she heard, so she signed on for the next work camp, which was in February 1992. Olive Jean led the team back to Coahoma.

Holly's experience in Mississippi was very special. Her job there was like no other she had ever had. The crew from Plymouth worked hard all day long, side by side with the future homeowners. They shared meals, cleanup, and the other daily chores. They met Mayor Jones and got to know the community, which calls itself "the little town with a million friends." They learned new skills and felt God's power in doing His work.

On Wednesday of that special week, Holly had a powerful spiritual experience. She was installing a tub in one of the Habitat houses. It was cold and progress was very slow. She had a lot of quiet time alone in that small bathroom. Suddenly the reality of what she was doing overcame her. Joyful tears streamed down her face as she realized that God was answering her prayer of many years. Seven years earlier, Holly had purchased an old house in Minneapolis, and she felt good about having her own roof over her head, so good that she immediately started praying, asking God to help her put a roof over another woman's head. Now here she was, on her knees, helping to build a house for Helen Jackson, her elderly mother, children, and a grandchild. Holly felt God's presence with her in a profound way at that moment of inspiration and revelation.

That experience and others in Coahoma had a transforming effect on Holly and the other women with her. Upon returning home, Holly increased her participation in the work camps. She returned to Mississippi, traveled to Columbia, South Carolina; went to Eagle Butte, South Dakota; and added workdays with her local affiliate, Twin Cities Habitat for Humanity. Meanwhile, Olive Jean, along with local residents of Coahoma, also incorporated Education for Coahoma, a nonprofit organization to seek to improve educational opportunities in the town.

Now Holly's enthusiasm matched Olive Jean's, and both were talking about it to everyone they knew, especially at church. Other women, including Alice Tuseth, were also deeply affected, and their enthusiasm spilled over and poured through Plymouth Church. A Habitat for Humanity committee was formed in the church to coordinate work with Twin Cities Habitat. Alice became the backbone in helping Plymouth Church develop a strong and growing partnership with Habitat for Humanity. These women, so affected by the blessing they received from building houses with Habitat, were now allowing the "more" to find its way into their house of worship.

A College Entrance Essay

The final story in this chapter is the most touching. It is about Mark Domaleski, a five-year volunteer for Lynchburg, Virginia, Habitat for Humanity. Mark's life was changed profoundly by his experiences with Habitat as revealed by his college entrance essay, which he wrote after volunteering with Lynchburg Habitat during the summer of 1993.

My work with Habitat for Humanity changed my outlook on life and reinforced my belief in the goodness of mankind. I gained a wealth of knowledge about building houses, but more importantly, I learned valuable lessons about myself and other people.

Imadu Cecil proved to be one of my best friends over the summer. Imadu, a seventeen-year-old, was helping to build his home and his neighbors' houses. Since we were the youngest on-site, we always seemed to be paired on the most undesirable jobs. Whenever someone needed a footing dug or a foundation tarred, we were automatically nominated for the task. However, it was through this hard work that we grew close. Whenever the sun got too hot, the wheelbarrow got too heavy, or the chore seemed impossible to complete, he supported me, and I him. We both took pride in our work. Whenever we started a task, we made sure it was done to perfection. I am looking forward to attending Imadu's house dedication. I will gain a feeling of satisfaction out of seeing a completed structure that I helped build. I am also glad that my friend will have a nice place to live.

Dr. David Harris, a retired surgeon, was also on-site every day. Rather than spend his free time fishing or playing golf, he has decided to make a difference in his community. Under his guidance, I learned how to hang dry wall and work a skill saw. Beyond that, he shared with me his knowledge and past experiences. Through his example, I learned that one can find fulfillment by giving of oneself. Through working with Dr. Harris and the many other retired people, I gained an immense respect for the elderly. I learned that seventeen- and seventy-year-old men have a lot in common, particularly when it comes to humor.

I have always prided myself with having empathy for those less fortunate than I. This experience provided me an opportunity to help and to get to know them. Pam and Arthur Brown will own one of the houses on which I worked. Through them, I realized how difficult it is for people to break the cycle of poverty. One of the most touching moments during the summer was when the Browns were told their application for a home had been approved. Their joy was contagious: for Pam and Arthur, it was a dream come true; for those present, it was the joy of helping make a dream come true.

When I decided to volunteer at Habitat, I never foresaw the personal growth that I would attain. Kevin Campbell, the Habitat director, and Mary Adams, the volunteer coordinator, became my role models. When I saw the good that they did and the satisfaction they gained from their work, I became committed to the organization. What started out as a part-time volunteer work became a full-time job, as reflected by my hours. This experience gave me direction for my future. I do not know what road I will follow in life, but at the end of my journey, I want to have made a contribution to the betterment of society.

Unfortunately, Mark's journey on this earth was not long. He drowned in a kayak mishap in late 1997. At the time of his death, he was working with AmeriCorps in Delaware.

Mark Domaleski's life was changed by his good work with Habitat for Humanity. And that changed life was a blessing to all who knew him. Now, through this book, his dedicated life of service will be known by thousands of additional people who will be inspired and impacted by his example.

From a young couple in Alaska to an older couple in Virginia, from a single mother in Kenya to a family in Sri Lanka, from a student in Ohio to a retired navy captain, hundreds of men and women from all walks of life have learned new skills, found new direction, and launched new careers after being a part of Habitat for Humanity.

CHAPTER 8

Inspiring Love
and Marriage

I still enjoy it when people find out that I met my husband at a
Habitat site. We thank God for bringing us together.
—ARDATH PHILLIPS, HABITAT VOLUNTEER

A couple meets. They fall in love and get married. It happens every day around the world. But what might surprise you is that it is also happening, with increasing frequency, to people in the ministry of Habitat for Humanity. In fact, it is happening so much that this wonderful fact of life certainly deserved being in a book that deals with "more than houses" in Habitat life.

I remember one day at devotions at our international headquarters in Americus that three couples who had met at Habitat announced their engagements—three in one day. I often say jokingly that we have these Habitat romances and marriages for people to get together and create "Habitots." Hundreds of these Habitots are running around various places on the globe. Some are even getting old enough to help build Habitat houses.

Holding Hands by Tennessee

One of the first Habitat marriages was that of Dan Roman and Sally Winter. Dan was an international partner working in Africa. Sally was a nurse in Ohio. In 1983, Linda and I led a walk from Americus to Indianapolis to celebrate the seventh anniversary of Habitat for Humanity. Sally was with us from the start. Dan, returning from Zaire, joined the walk in north Georgia, where we were camping in a grape vineyard out in the country just outside of Cartersville. Immediately, Linda noticed the potential for romance. Dan and Sally were made for each other in her opinion. Sally had recently lost her fiancé to cancer, and Dan had returned home to learn that his longtime girlfriend had married someone else. As we walked steadily northward toward Indianapolis, the romance blossomed. Linda and I began to speculate—Will they hold hands by Tennessee?

Yes, they did—in northern Tennessee.

Will they walk arm in arm in Kentucky?

Yes, they did—in central Kentucky!

By the time we reached southern Indiana, she was sitting in his lap during breaks. A year later, they married in her hometown of Akron, Ohio. Linda attended the wedding.

Dan and Sally worked for Habitat in Africa, Dan serving as national director of Zaire from mid-1990 to 1991, followed by two years in Ghana as West Central Africa regional director. Then a serious illness of their son, David, caused them to return to the States, and now they live and work in Mesa, Arizona. Dan and Sally no longer work for Habitat, but they remain supportive of the ministry. God has blessed their marriage with four Habitots.

Wedding on the Walk

Fred and Ellen Schippert are another couple who met, fell in love, and married as a result of their involvement with Habitat for Humanity. Fred was a retired mathematics supervisor for the Detroit public schools. He came to Americus to volunteer in 1984. His first wife had died ten years earlier. Ellen was an architect from Youngstown, Ohio. She came to Americus to volunteer in 1985. The two of them met while working at headquarters, and their friendship evolved into a deep love for one another. In 1986, Linda and I organized another walk, this one from Americus to Kansas City to observe our tenth anniversary as a ministry.

Fred signed on for the whole walk. Ellen joined us in Springfield, Missouri. The two of them had decided to get married on the walk. As we discussed the idea, we had a lot of fun talking about finding a preacher who could walk backward and perform the wedding ceremony on the road.

It was finally decided that they would get married in a church at the end of the day on September 16, 1986. Walkers picked wildflowers along the highway and decorated a Methodist church in Harrisonville, Missouri, some twenty-five miles south of Kansas City.

The wedding ceremony was conducted by three fellow walkers: Mel West, a retired United Methodist pastor from Columbia, Missouri; Korabanda Azariah, founder of Habitat for Humanity in Khamman, India; and the Right Reverend Benjamin Ogwal, an Anglican bishop from Gulu, Uganda. A beautiful and very significant part of the wedding ceremony was a recitation of the Lord's Prayer, led by Bishop Ogwal and spoken in thirteen different languages. I was privileged to serve as best man for Fred. After the wedding, a reception was held in a nearby Catholic church, and the next morning the group walked on into Kansas City for the celebration. It was a joyous time for everyone, but Fred and Ellen were most joyous of all.

After their time as volunteers in Americus, they worked together in 1986 in founding Metro Detroit Habitat for Humanity. Then, in 1988, they returned to Americus to enter training to become international partners in Mexico. They served first in Anahuac and then, in 1989, moved to Mexico City. Fred and Ellen continued to serve in Mexico City until early 1992, at which time they moved to Palms, Michigan, and began working with Mid Thumb Habitat for Humanity. Fred was soon president, and Ellen was on the board. Later, Ellen became heavily involved with the Michigan Habitat state organization, serving as architect for all statewide affiliates. Fred and Ellen have been married for more than twelve years. They are still very much in love and still very much committed to Habitat for Humanity. They participated in the 1987 Jimmy Carter blitz-build in Charlotte, North Carolina. They also journeyed to the Philippines for the Jimmy Carter work project in 1999 and say that they are excited about their continuing involvement in this ministry of hammers and nails, and love and marriage.

A Year to Decide

Michelle Gordon and Peter Dalva met in July 1991 on Habitat's Fifteenth Anniversary Blitz-Build. That unique venture involved fifteen work teams starting from various points around the United States and building their way over fifteen weeks toward Columbus, Ohio, for the fifteenth anniversary celebration. About fifteen hundred houses were worked on or completed during that busy summer. Michelle and Peter were on the work team that started in Gainesville, Florida. That's where they met.

Peter was a recent graduate of the University of Michigan, and Michelle was from the University of Wisconsin-Whitewater, where she had been active with the Habitat for Humanity campus chapter. After spotting a Habitat notice on a bulletin board at college, Peter signed up for the summer of "traveling construction." He remarked, "I couldn't think of a better way to spend my summer."

"When first seeing Peter, I figured him for a real jerk," said Michelle. After three weeks of traveling together on the same construction team, though, the two became close friends, spending nearly every waking moment together.

One stop along the fifteenth-anniversary traveling work camp was at LaGrange, Georgia. Each evening after construction had come to a close, Peter would borrow a car from his host family and take Michelle on dates. It was on these dates that the friendship blossomed into romance.

The summer ended, and Michelle and Peter parted ways in Columbus, Ohio. There were no commitments or promises. Michelle returned to Americus to work in the headquarters, and Peter returned home to New York.

Within four days, the two were writing letters. Then, in April 1992, Peter came to Americus as a volunteer in construction. Michelle bluntly told him upon his arrival, "You've got a year to decide whether you're going to marry me." Just about a year later, Peter proposed, and Michelle accepted. Incidentally, the folks in LaGrange, Georgia, insisted on throwing the wedding, and on April 23, 1994, Peter and Michelle were married there. Both of them continue to work for Habitat for Humanity International, Peter in the operations department, and Michelle in the human resources department.

China, Americus, and Ted

I am proud that I had a small hand in finding a mate for my good friend, Ted Swisher, Habitat's director of all U.S. affiliates. In 1987, I was on a speaking tour in New Mexico. One evening, I was speaking at New Mexico State University in Las Cruces. Following my speech, people came up to greet and talk to me. In the crowd was a man and wife and their beautiful daughter, Lisa Verploegh. I learned that Lisa had applied to be an English teacher in China but was having trouble getting a visa. I began to joke with her, asking why she would want to work in China when she could be a Habitat volunteer in Americus, Georgia. Having learned that she was quite creative, I suggested that she work with our creative services department in the headquarters. She was not sure at that moment, but the seed was planted. In May 1987, she arrived in Americus to work. Whatever I said, it worked—China became a distant memory.

Before too long, Lisa and Ted met and, over the next several months, fell in love. On November 22, 1987, they were married in the Lee Council House, just one block from our house on East Church Street. After the wedding, there was fun and games, including a soccer match at a nearby athletic field. Ted and Lisa went to Australia in 1989 to launch Habitat's work in that country. Their first child, Benjamin, was born there in 1990. After moving back to Americus, Miriam Grace was born in 1994.

Love Abroad

Often at Habitat for Humanity, people find love outside the boundaries of the United States. After all, at the risk of sounding sentimental, love has no boundaries.

George Keller had to travel abroad to find his wife. He came to Americus in 1992. George worked at various jobs over several months, most related to construction. Then, in 1993, he entered training to be an international partner in Fiji.

George and Alumeci Keller pose for pictures after the joyous ceremony.

During his work in Fiji, George met Alumeci-Y-Kalou. He decided that she was the love of his life. Fortunately, she agreed.

George returned to Americus without Alumeci, but he was determined to raise the money and overcome the obstacles to bring her to Georgia to be his bride. Finally, on January 24, 1994, Alumeci arrived in the States. George was in bliss. The two of them were married at the Habitat Center in Americus on February 13, 1994.

Both of them were actively involved in the ministry in Americus until early 1999. Alumeci worked in our print shop, and George headed our tour program, which greets and gives tours to the ever-increasing hundreds of visitors who come to see our headquarters and the work being done by Habitat for Humanity in southwest Georgia. In March 1999, they returned to Fiji.

Wayne Nelson also met and fell in love with a woman from another country, but their romance blossomed in Americus. Wayne started his work with Habitat for Humanity in Michigan in 1982 at the Lake County affiliate. He joined our staff in Americus in 1987. Over the years he served in several capacities, primarily related to construction and environmental concerns.

Gera Goncalves began her Habitat career in her native Brazil, in the city of Belo Horizonte. In 1990, she became an international partner, serving first in Nicaragua and then in the Philippines. In 1994, Gera came to Americus and met Wayne.

She and Wayne were married on September 3, 1994, in a beautiful wedding in Midland, Michigan. And they now have two Habitots. Wayne continues to work at Habitat International in the construction and environmental resources department.

Americus—"Heartland" for Volunteers

Often volunteers who come to Americus for training leave with much more than they ever thought they'd find, and I'm proud to say that has included love and marriage for many special couples.

Claire Williams and Steve Martindale found each other through Habitat in Americus. Claire came from Austin, Texas, to volunteer at Habitat headquarters in early 1982. She stayed for two years, helping to coordinate other volunteers and host work camps and, along with coworker Nancy Braud, served as a codirector of the seven-hundred-mile walk Linda and I led to Indianapolis in the summer of 1983. After a few years back in Austin, working with the local affiliate, Claire returned to Americus to go on staff with the affiliate department. Over the next five years, we kept her very busy working in a variety of assignments, such as helping to organize the one-thousand-mile walk to Kansas City, Missouri, and also to coordinate and nurture Habitat affiliates around the United States, particularly in the southeast.

Meanwhile, Steve Martindale got interested in Habitat through the Brethren Volunteer Service in Roanoke, Virginia. In 1987, he came to Americus as a volunteer. He and Claire met and, after a courtship of several months, married on July 6, 1991. They have been a solid Habitat love story ever since.

In 1991, Claire became the first regional director of the Habitat East Region, which includes Virginia, West Virginia, Maryland, and the District of Columbia. She served in that role until 1996. Currently, Claire and Steve are active with Central Valley Habitat for Humanity, where they live in Bridgewater, Virginia.

Carolyn Ross, a native of Willard, Missouri, came to Americus as a volunteer in 1985. Pete Talboys of Clarence City, New York, arrived a year later. But it didn't take them long to notice each other. Soon they were romancing each other too. Wedding bells rang for them on April 4, 1987. From 1989 to 1995, Carolyn was director of the Heartland Region of Habitat for Humanity, headquartered in Springfield, Missouri. Pete and Carolyn have one Habitot. And even though no longer employed by Habitat, Carolyn and Pete remain supportive of the ministry that gave them each other.

Holly Sawyer learned about Habitat for Humanity from her father, Warren, who served on the Habitat International board for several years. Holly decided to be a volunteer in Americus. She arrived in October 1987. Bill Branner of Beaverton, Oregon, decided to be a volunteer in Americus and arrived at the headquarters in March 1987. They met, fell in love, and were married on August 26, 1989, in Swampscott, Massachusetts. Soon thereafter, they entered training to be international partners. They served faithfully for three years in Guatemala then established Habitat for Humanity in Bogota, Colombia. They have two Habitots, and the family currently lives in Lynn, Massachusetts.

Employee Fraternization

One of the most recent Habitat marriages is that of Steve Messinetti and April McGinnis. The two met at Habitat headquarters in January 1994 , their first day of work in the campus chapters department. Steve was taken with April from the start. After three weeks of training, they were assigned

Steve and April Messinetti on their wedding day.

various regions of the United States. April's first assignment was in North Carolina. Knowing that April would need a ride, Steve offered to drive her to North Carolina. On the trip, he told her of his interest in her. April, however, didn't reciprocate the feeling, and upon further deliberation and contemplation, they both agreed that they were incompatible and should never date.

Over the next two years, however, they became close friends and continued working in the same department. Eventually, April realized she was developing feelings beyond friendship for Steve, after all. The two started to date—quietly, to avoid rumors of employee fraternization. Then in 1997, the week before Christmas, Steve asked April to marry him. They were married in Oregon on October 4, 1998, and they continue to "fraternize" in the same campus chapters department.

Marriage on a Roof

So many wonderful love stories have happened because of Habitat over the years, and not just through an Americus Habitat headquarters connection. Many romances start while volunteers are hammering side by side here, there, and everywhere. All across the United States and abroad, romances have started, some blossoming in very unusual places, like, for example, on a roof!

Vic Fasolino, from Wrightstown, New Jersey, became involved with Habitat in 1991, when he participated in a traveling work camp that ended in Sioux Falls, South Dakota. Following that trip, Vic not only began serving on the board of the Habitat affiliate in Trenton, New Jersey, he became involved in the international work of Habitat. Later in 1991 and again in 1993, Vic traveled to Guatemala to build Habitat houses. In 1995, he took part in a Global Village work camp in Kenya. Vic had quickly become a dedicated volunteer and a Global Habitat partner. But the highlight of Vic's overseas Habitat experience was still to come.

Living in Seattle, Washington, Lora Fowler started working with the local Habitat affiliate in 1994. Volunteering on-site quickly progressed into serving on the board and chairing several committees. In the summer of 1996, Lora decided to volunteer at the Jimmy Carter work project in Hungary, her first international Habitat build. There she met Vic Fasolino, on a roof. Over the symphony of hammers, a friendship developed as they installed roofing together that summer. In the months that followed, friendship transformed into deep affection and love. Vic and Lora decided to do something bold. Already they knew they wanted to work on the next Carter project, which was scheduled for eastern Kentucky and Tennessee in June 1997. "Why not get married on the roof of the house we build in that project?" they decided. "We met on a roof. We should get married on a roof!"

Vic and Lora labored diligently at the Indian Creek site in Virgie, Kentucky to build a house for Janelle White and her daughter, Destiny Hager. Then, on the Friday at the end of the blitz week, they were married while standing in the "valley of the roof" of the newly completed house. I was one of the privileged people who witnessed the unusual wedding ceremony. Hundreds more stood shoulder to shoulder in the front yard.

Vic wore a tuxedo with a few extra touches (the cummerbund was a Habitat nail apron). Lora arrived in a traditional white wedding

Vic and Lora Fasolino wearing rooftop smiles at their most unusual ceremony. Photo by Charles Bertram.

dress with a long train. Vic went up the ladder first, and Lora followed. What a magnificent sight! The train totally covered the ladder as she reached the top. When she joined Vic on the roof, a fellow volunteer lifted the train up behind her.

The brief ceremony was performed by the Reverend Jon Ford of Pikeville's First Christian Church. As Vic and Lora were pronounced as husband and wife, he lifted the veil and kissed her. In the excitement of the moment, Vic temporarily lost his balance, quickly regaining it to all our gasps—thankfully—and finished that kiss.

A week later I was in Detroit for another blitz-build. Vic and Lora were building away, on their honeymoon. A year later, I was in Seattle on my annual tour of some of the Building on Faith sites. Again, I spotted Vic and Lora. They told me that they had participated in ten Habitat builds since their wedding. Early 1999 found Vic and Lora in the Philippines helping with the extensive preparations for the Jimmy Carter blitz-build there in March.

Roofs, Porches, and Floods

Another couple had their first workday on a roof of a Habitat house in Knoxville, Tennessee. Six weeks after meeting, Paula Blankenship and John Smartt Jr. worked together on a roof. Later in the summer, they completed the vinyl siding on several houses in Rugby, Tennessee, during the 1997 Jimmy Carter work project. On January 2, 1998, Paula and John were married. John continues to work as a crew leader with Knoxville Habitat for Humanity. Paula joins him several weekends each month. John says that Paula's favorite activities are putting on roofing shingles and installing vinyl siding. "In more ways than one," he said, "Habitat has changed my life." He says that he has a fulfilling volunteer life and part-time professional life that gives him the best of everything. And he is grateful to Habitat for helping him find common interests with "a most uncommon, loving woman."

David Baldwin met his future wife not on a roof but on a porch. Jill Shelley was on-site volunteering for the first time with Habitat for Humanity in Manhattan, Kansas. Dave had been volunteering for several weeks. The two of them ended up painting the porch together and talking. They discovered that they had gone to school together at Bethany College in 1977 but had never met. Several weeks after that first meeting, Dave called Jill and asked her to help him put up a fence at the Habitat house. Two months later, they were married.

In December 1996, Dave and Jill moved to Lawrence, Kansas. Later the next year, Dave was hired as volunteer manager of Lawrence Habitat for Humanity. Now Jill frequently is one of the volunteers "managed" by Dave.

Sometimes, romance buds with Habitat homeowners, resulting in wedding bells. That was the case with Pam Noah of Colfax, Iowa. Pam had been putting in sweat-equity hours on her Habitat house for over a year in addition to raising three daughters as a single parent. At that time she was living in a dilapidated two-bedroom trailer. But then a short in the trailer's wiring forced the family to move. Pam stored her belongings in her parent's garage and "camped out" in her mom's living room.

A month later a flood came to the town, and Pam's parents' house was hit hard. Pam lost everything she had stored there. The whole town was overwhelmed with flood recovery, and progress on Pam's house slowed considerably. The local Habitat affiliate decided to work in the evenings, two or three nights a week. That turn of events caused Pam to meet James Braunschweig, a Maytag engineer and volunteer house crew leader.

"I had heard his name a lot," she said, "but I had never met him until after the flood when the night crews started." Hearing about him turned into liking him, which turned into love on both sides. Pam and James were married in February 1994, just two months after Pam and her daughters moved into their new three-bedroom Habitat house. Since their marriage, Pam and James have remained involved with the local Habitat affiliate, Jasper County Habitat for Humanity. Jill works at a local jewelry store and is doing some public relations work for the affiliate. James is still working for Maytag and was the site supervisor for the affiliate's fourth house in early 1999.

In Denton, Texas, Pam Davila and Larry Monroe were both members of St. Andrew Presbyterian Church. In 1996, their congregation and three other Presbyterian churches decided to sponsor a Habitat house. Plans were made and money raised so construction could begin in the fall of that year. Pam and Larry decided separately to participate in the building of the house. They had met at church but were only acquaintances. Each had been married before but were now single.

Pam brought her camera each week and took pictures of the house as it was going up and of the volunteers. As time went along, Pam noticed that Larry was in a lot of her pictures! Other volunteers began to tease Pam and Larry about how they always seemed to be installing siding on the same side of the house or painting in the same room. Maybe they were. At Christmastime, Larry asked Pam to marry him, and she said yes. In January, all of the regular volunteers were given invitations at the job site to attend Pam and Larry's wedding. The house was completed in March 1997. But there was a joyous reunion in May when the volunteers attended Pam and Larry's wedding at the church. The volunteers gave them a matching set of pink and blue hammers in a specially designed box.

A Habitat Romance

In Indiana, Jane Blaeholder got something she wasn't expecting when she began working for a new Habitat house; she also got a new husband. Jane was a single parent with three children, and she was working two waitress jobs when she was selected to be a partner family with the Greater Muncie Indiana Habitat for Humanity affiliate. One of her jobs was at a busy truck-stop, the other at one of the fanciest restaurants in town. She talked to everyone she met about Habitat, attempting to get new volunteers to help build her Habitat home as well as others. She wasn't having any luck until the young man who handled the heavy-truck mechanical repairs at the truckstop asked her to tell him about it. Soon she had a volunteer.

Alan Bowden was the "hardest-working man she had ever met," said Jane later, and she began to realize he was also "sweet and nice." They worked every day on the home which was completed sooner than any other home the affiliate had ever built beyond a blitz-build. The only time he took away from the site was to be Jane's assistant Little League coach for Cub Scout baseball. The house was built, and so was their marriage and future life together.

Love and a Birdhouse

Atlanta Habitat for Humanity hosts an annual fund-raiser called Birdhouse Artfest, where birdhouses are designed and painted by famous artists in the southeastern United States and then auctioned off for Habitat's benefit. Susan Shaver had volunteered at these Birdhouse Artfests since 1990. She had been active with Atlanta Habitat for Humanity since 1989, after graduating from Emory Law School in Atlanta. She was involved in construction, as a member of the family selection committee, and as an attorney who volunteered for closings on Habitat houses. In 1993, she was enchanted with a beautiful copper-gourd birdhouse, so she bid on it and purchased it.

The following year she was introduced to a dapper young gentleman, Scott Wiederholt, whom she immediately recognized as the artist who had created the birdhouse she had purchased. Love blossomed in the coming months, and wedding bells rang in May 1998. The couple continues to be involved with Atlanta Habitat for Humanity.

Phone Number on Siding

Ardath and Roger Phillips are another couple who met and fell in love on a Habitat work site. Their rendezvous was at a building site of Habitat for Humanity of Greater Canton, Ohio. Ardath tells the exciting story of how they met, fell in love, and married. "Roger and I met while working on a housing site in October 1994. He worked for a company that had donated materials and labor for siding the house. I had volunteered to help with the work. He taught me how to hang siding. As we worked together, we talked all morning. At the end of the day, he took my phone number on the only thing handy—a piece of siding. We dated all winter and were engaged in the spring."

That summer was the seventh-anniversary housing blitz in the same neighborhood where Roger and Ardath met and where they again were volunteering. While talking to a fellow volunteer, Ardath told how she met her fiancé on a work site. "The story got passed around, and by the next day there was a pile of siding scraps at the base of a tree in the yard with phone numbers," Ardath said, laughing. "People began to make wedding suggestions: a twenty-one nail gun salute and a recessional of crossed hammers. We felt both honored and very special. We were also lucky to find a minister at the blitz to marry us."

The Reverend Rich Plant officiated at their wedding at Trinity United Church of Christ the following June. Ardath says she almost cried when he talked about the unexpected benefits of the ministry of Habitat for Humanity! "Today, I still enjoy it when people find out that I met my husband at a Habitat site. We thank God for bringing us together."

Made in Heaven and Habitat

Another Habitat love story happened in Daytona Beach, Florida. Jim and Robyn Seidel first met in September 1995, when Jim came to work for Halifax Habitat for Humanity as the general contractor–construction supervisor for a blitz-build. Jim is one of the most respected builders in the Daytona Beach area. Robyn, Halifax's administrative assistant, has a magnetic personality displayed every time a caller dials the Habitat office. Robyn and Jim had their first "official" date in May 1996, when they attended Halifax Habitat for Humanity's ten-year anniversary party, at which I was the guest speaker.

"Halifax Habitat for Humanity had their wedding of the decade on Saturday, June 28, 1997," stated Jane Walters, executive director of Halifax Habitat. "Robyn Eileen Seidel and James Edward Cary were united in marriage at Westminster by the Sea Presbyterian Church. The Reverend Jeffrey Sumner officiated as the bride and groom's family members enjoyed a beautiful ceremony in the famous Rose Chapel. Robyn's daughter, ten-year-old Jessie, dignified the occasion as maid of honor. Jim's grandson, Scott Williams, was best man. And Robyn's dad did not let his inability to walk hinder him from giving his daughter away. Family members from both the groom and bride's side came from all parts of the country. It truly was a wonderful occasion. And without a doubt, the marriage was a match made in heaven."

"To show how dedicated both Jim and Robyn are," added Jane, "on Monday, June 30, Jim was back on the job with our many houses under construction and Robyn was back at the switch in the office. They did take time off in July, however, for a fabulous honeymoon cruise. Robyn and Jim's marriage has provided a moving, dynamic force for Halifax Habitat's great plans for the twenty-first century."

Building a New Life Together

One of the most touching Habitat love stories is that of Mac and Doris Young of Augusta, West Virginia. Mac, a recently retired electrical engineer with some woodworking and home-building experience, lost his wife to cancer in December 1994. Mac and his late wife had retired to Augusta and had helped build their retirement home on thirty scenic acres of ridge land. In the spring of 1995, Mac answered the call for volunteers willing to help organize a new Habitat for Humanity affiliate in Hampshire County, West Virginia. He decided that building houses for people in need was exactly what he needed to begin building a new life for himself. Mac served on the steering committee, organized the building committee, and became a member of the first board of directors in the fall of 1995. At the December 1995 board meeting, Doris L. Chule was introduced as a prospective new board member. Doris, a retired senior accountant, had lost her husband to cancer in December 1994. She and her late husband had built their retirement home in Hampshire County about the same time as Mac and his late wife.

Mac and Doris had never met, even though both of them had made many trips to the same cancer treatment center during 1994.

Mac and Doris were appointed to the new church-relations committee formed at the December meeting, and Doris was elected to the board at the January 1996 meeting, thus creating two monthly opportunities for contact. Furthermore, they found that they were both attending the monthly meetings of another local organization. Soon personal meetings were taking place that had nothing to do with Habitat for Humanity or any other organization! Rather quickly, Mac and Doris decided that it was God's work in bringing them together. They were perfectly matched in all ways and could see no reason to delay getting married. John Franklin, pastor of the Romney Presbyterian Church and a member of the Habitat board of directors, married them at Mac's home on April 15, 1996, with a small group of family members present.

Mac and Doris wanted to celebrate their wedding with a delayed reception for a larger group of their families and many friends in Habitat and the larger community, but they did not need or want wedding presents. After selling Doris's home and combining their worldly goods in Mac's larger home, they found duplicates and sometimes triplicates of nearly everything. The work of God again became evident as they struck on the idea of turning the wedding reception into a Habitat celebration fund-raiser. Dan Stiltber, pastor of the Augusta Church of Christ and also a member of the board, offered the use of his large fellowship hall and kitchen facilities, and on May 30, 1996, some one hundred people gathered at a dinner to celebrate both the wedding and Habitat for Humanity. This event raised seventeen hundred dollars toward building the affiliate's first house.

Mac and Doris remain active with Habitat for Humanity. Doris is treasurer and serves on the fund-raising committee. Mac is cochairman of the building committee. The newlyweds are enthusiastic campers who enjoy traveling in their motor home when they are able to tear themselves away from Habitat and church commitments.

Love and marriage are a significant part of Habitat. It brings me great joy to know that this ministry has brought so many people together who have found happiness in new lives together as they have worked to bring happiness to others. More than houses? Indeed! And may the wedding bells keep ringing as the hammers resound.

CHAPTER 9

Creating Friendships

Habitat for Humanity—What does it mean to me?
A time for building a home, don't you see.
There's much more to a home than four walls and a roof above.
There are volunteers, supplies, and plenty of love.

A time to laugh, a time to share.
A time to show a family how much we really care.

Friendships will begin right from the start.
Some of us will be pals, deep within our hearts.

As for me and my crew-member role,
I truly can say Habitat for Humanity has touched my soul.

I wish that everyone could enjoy this great experience that is near.
And maybe by chance, we'll see each other next year.

So the next time you put on your work boots and gloves,
Remember, my friend, the true meaning of love.
—KIMBERLY KEITH, AT THE HOUSTON BLITZ-BUILD, JUNE 1998

Building friendships is a huge part of the "more than houses" venture of Habitat for Humanity. I find myself smiling as I begin writing this chapter and remembering the hundreds of friends I have made over the years in this work. Let me share with you just a few of the special people who have become friends through this hammering ministry and who have been pivotal to its success.

First, and most important of all, there is Clarence Jordan, with whom I started this house-building enterprise at Koinonia Farm in 1969. Of course, it was not called Habitat for Humanity back then. The name we used was simply Partnership Housing.

Clarence, his wife, Florence, and another couple, Martin and Mabel England, had launched the Koinonia ministry near Americus in rural southwest Georgia in 1942. They wanted to develop a community patterned after the early church where everyone shared all possessions in common. And they desired a totally nonracial fellowship where blacks and whites would live and work together as equals. Over the years, the dream was realized. A thriving community emerged, and they supported themselves by farming and selling their produce.

The surrounding white community did not appreciate this integrated group in their midst. So in the late fifties, violence erupted against the farm. There were numerous shootings, bombings, burnings, and beating of Koinonia people when they went to town. An economic boycott was set up to drive the community out of the county. Many people at Koinonia could not take the intense pressure, so they left. Only a small remnant remained.

This small remaining group survived by their wits and hard work. Clarence Jordan, a first-rate Bible scholar with a Ph.D. in the Greek New Testament, spoke widely all across the United States. Fees from those speaking engagements helped his family and others in the community to survive. Clarence also started to write, primarily translating the New Testament into his "Cotton Patch" gospel. In that translation, Clarence brought Jesus to Georgia. Jesus was born in Gainesville and crucified in Atlanta. A series of these books was published: The Cotton Patch Gospel of Luke and Acts, the Cotton Patch version of Paul's Epistles, and others. Before his untimely death on October 29, 1969, Clarence had translated nearly all of the New Testament into his Cotton Patch version.

Income from these books helped sustain the community. Also, Clarence launched a small enterprise of selling pecans and pecan products by mail. The advertising slogan for the new business was, "Help us ship the nuts out of Georgia!"

Linda and I arrived at Koinonia with our two young children, Chris and Kim, in early December 1965. It was a time of great transition for us. We had never heard of Clarence Jordan. Very quickly, though, we got to know this extraordinary man. I felt, literally, that God had guided us to him. Our visit of two hours lasted for a month.

During that time, I helped pack pecans with Clarence. He and I milked the community cow. We took walks and shared many meals. He introduced me to a Jesus I had not known. We talked about an uncertain future that lay ahead for Linda and me. We read Scripture and prayed together. We bonded. A lifetime friendship was formed.

Our family left Koinonia and my new friend at the end of that December. Over the next two and a half years, I was engaged in other pursuits, but I stayed in touch with Clarence. Then, in mid-1968, Linda and I, along with our children, expanded by one with the birth of Faith the previous May, moved back to Koinonia to work full time with Clarence and the others at Koinonia.

Partnership Housing was launched by starting one house for one needy family. While that house was under construction, on a crisp, cool day in October, Clarence died of a sudden heart attack. One of the great honors of my life was conducting his funeral.

Clarence was a marvelous friend. He had also become my spiritual mentor. His impact on my life has truly been incredible and awesomely profound. I think of the old Christian hymn "What a Friend We Have in Jesus" when I reflect on my deep friendship with Clarence Jordan. What a friend I had in Clarence. Even now, he continues to impact my thinking and my whole life powerfully.

Other friendships have developed over the years. Ted Swisher, a young idealist from Pittsburgh with a Princeton degree in hand, showed up at Koinonia Farm in 1970. He and I became coworkers and friends. In 1983, Ted became the first U.S. director of Habitat for Humanity. In 1989, Ted and his new wife, Lisa Verplough, went to launch Habitat's ministry in Australia. Three years later they returned to Americus, and Ted became director of the ambitious Sumter County Initiative to eliminate all poverty housing in our town and county. Under his expert guidance, the program got off to a grand start. In 1997, Ted once again was named director of all U.S. Habitat work. Ted is still on the job, faithful and able, overseeing more than fifteen hundred affiliates in all fifty states. I feel so blessed to have such a wonderful colleague and friend.

George Peagler, an outstanding local attorney, and Russell Thomas Jr., mayor of Americus, are two other cherished friends who have worked tirelessly with the local initiative to end substandard housing in Americus and Sumter County. I have a great love and appreciation for both of these men.

Clive Rainey became our first Habitat volunteer in 1977. Since then, he has served in a number of roles in this ministry, including director of our work in Africa and a senior officer in our development department, as mentioned in chapter 4. In 1999, he took on the exciting task of coordinating the growing number of towns, cities, and counties that have decided to "set a date" for ending poverty housing in their local area through the program, the Twenty-first Century Challenge. Clive is a personal friend in addition to being a dedicated and effective partner in the ministry of Habitat for Humanity.

Margo Mitchell, a new homeowner, shares a tender moment with Habitat volunteer Marilyn Hult in Americus, Georgia, during the Easter blitz-build in 1998. Photo by Kim McDonald.

Jacob Battle, a local businessman and also a preacher, became my friend nearly a decade ago. He is a real friend who constantly calls to check up on us and take me fishing at every opportunity. That's not very often, but Jacob knows that I need a break every now and then. Jacob also is a perpetual encourager of the Habitat ministry. He even has a collection box at the entrance to his store for his customers to contribute to the work of Habitat.

In Africa, when our family was there to test this concept of house building in Zaire (now the Democratic Republic of the Congo), I developed profound friendships with many people but especially with Pastor Lokoni, Mbomba, and Mompongo. I wrote about each of these special men in my first book, *Bokotola.* They supported and encouraged me and the blossoming housing work, called at that time simply the housing project, in truly awesome ways.

While still in Africa, I started the manuscript for *Bokotola.* Upon returning home, I collaborated with a remarkable woman from Salem, New Jersey,

Diane Scott, to complete and have it published. Diane and her husband, Victor, had long been supporters of our emerging ministry, going all the way back to our days at Koinonia. Diane and I became friends in the process of writing that book, and over the years we have exchanged literally hundreds of letters.

Bill Clarke is an electrical contractor in Canton, Ohio. He was involved in Koinonia and then signed on with Habitat for Humanity when it got started. After Linda and I returned from Africa, we asked our sponsor, the division of overseas ministries of the Christian Church (Disciples of Christ) to support us for a year so we could start Habitat for Humanity. The division gladly did so. As we approached the end of that year, however, I realized that we still needed support for a while longer. I asked Bill Clarke to support my family for another year. He agreed to do so. That boost was the last push that was needed to launch the ministry of Habitat for Humanity. Bill also served faithfully on the board of directors of Habitat for Humanity International and generously contributed to the work every year. He and I became fast friends, and I developed a very deep love for him and for his family.

Jimmy Carter, of course, is well known all over the United States and around the world for his tireless work with Habitat for Humanity and for his many other worthwhile activities. I recruited him to join us in this ministry in 1984, and he has remained very committed in the years since. I have worked with President Carter at every one of his celebrated blitz-builds since 1984, and Linda has joined me on most of them. Both of us have an enormous love and appreciation for both President Carter and his wife, Rosalynn. President Carter has proven himself to be a faithful friend in countless ways over the years. He not only affirms and supports the work of Habitat for Humanity in numerous ways, but he also demonstrates his friendship in many very personal ways.

I remember talking to President Carter on June 30, 1998. We were discussing some matters relating to a forthcoming Habitat event. At the end of the conversation, I mentioned that our youngest daughter, Georgia, had given birth that day to our first granddaughter.

"What's her name?" he inquired.

"Sophie Davina."

A couple of days later, I received a nice letter addressed to Sophie, welcoming her into our family and into the world.

On January 3, 1999, Linda and I were at Sunday school at Maranatha Baptist Church in Plains, where we are members and where President

Carter teaches Sunday school. After making a few opening remarks, he walked over to us and asked, "Today is your birthday, isn't it?"

Former President Jimmy Carter (right) and Millard Fuller (left) pause at the 1998 Jimmy Carter work project to "upgrade" their equipment, courtesy of the retired chairman and chief executive office of The Stanley Works, Dick Ayres (center). Photo by Robert Baker.

"Yes."

"Then I think we should sing 'Happy Birthday' to you." And the class obliged.

If he hears there is an illness or some problem in our family, he will call or in some way let us know of his concern. I remember one time he called me during an extremely hard time in my life. He opened the conversation by exclaiming, "We love you." I was very touched by his obvious caring and deep friendship.

LeRoy Troyer has also proven to be an incredible friend. I had met him a few years earlier at a Mennonite Economic Development Association meeting in Cedar Rapids, Iowa. I recruited him for the international board. He signed on to work on the emerging Jimmy Carter blitz-builds. In 1986, he was the house leader on Jimmy Carter's house at the blitz-build in Chicago. He and I quickly became friends, and so did LeRoy and President Carter. He became house leader for the Jimmy Carter house on every Jimmy Carter blitz-build from that time forward. A bit later, his brother Lloyd joined the annual blitz-builds. Then, in the early nineties, when we needed more space at our headquarters in Americus, LeRoy designed the remodeling of an old building in downtown Americus that was to become our new headquarters. His brother Lloyd agreed to serve as the supervisor of the construction.

That facility was completed and dedicated in 1995. It continues to serve us well to this day. You are likely to find Lloyd and his wife, Loretta, helping out in the mailroom in the winter months. LeRoy is also an occasional visitor to Americus and a tireless worker with his local Habitat affiliate in northern Indiana.

The above, of course, are only a few of the strong friendships Habitat has given me over these years, but it shows how building friendships is such an integral part of the ministry of Habitat for Humanity. My life has certainly been enriched beyond measure over the years in the building of friendships so rich and so deep.

To work side by side in a cause that helps people so profoundly and so instantly is a natural place for forging friendships. And better yet, so many of these friendships wonderfully cross gender, racial, age, and other lines.

Across Age and Gender Lines

Evelyn Henville, executive director of Habitat for Humanity-Dekalb, in the metro Atlanta area, tells about a rich friendship that developed across age and gender lines. "In the spring of 1996 while working on a new house, I had the responsibility of finding lunch for the workers," she said. "Throughout the morning, fourteen-year-old Michael was ever-present, working diligently right next to me. Our conversations, as we worked, centered around his hobbies and mine, his likes and dislikes about school, and his great desire to get his driver's license. I never asked him where he lived or with whom he had come to the work site. I knew he was from the area, but what I did not know was that he lived right next-door and that his mom was Gladys Kinchen, a Habitat-Dekalb homeowner, whom I knew very well.

"When it was time for lunch, I said to Michael, 'I am responsible for getting lunch for the volunteers. Would you like to help me?' He readily agreed. 'Michael,' I said, 'I need a telephone book and a telephone to find a sandwich shop. Do you know where there is a public telephone?' He replied, 'There are no public telephones around here, but I know where there is a phone. Come with me.'

"Michael proceeded up the front steps of the house next-door. That was the Habitat house that had been built two years earlier and was owned by Gladys Kinchen. When he got to the top step, he sat down and began to take his shoes off," recalled Evelyn.

"As he did so, he said to me, 'Take your shoes off.'

"I did as I was told. Michael then started to open the front door. I asked, 'Do you live here?'

"'Yes,' he responded as he opened the door.

"With one hand on the doorknob and the other hand stretched royally to the ceiling, Michael curtsied with much grace, flair, pride, delight, happiness, and dignity. He then exclaimed, 'Nice, huh?!' The home was immaculately clean and beautifully decorated, and the pride shone on Michael's face.

"To this day, Michael and I remain the best of friends," said Evelyn. "He is seventeen now, and he and his mom always find time to help me mail the Christmas cards and Habitat newsletter."

A Bucket of Trash

Charlie Smith, vice president of Highland Rim Habitat for Humanity, Tullahoma, Tennessee, shared about a special friendship he built with a little girl. "Amanda was no more than five years old. One of five children in the Kochans family, she was eager to do her part so that her mommy and daddy would have their very own home. Later she would show me her room that she was so proud of. But for now she was helping the whole family pick up trash around the workplace," Charlie said. "Here she came with a bucket as big as she was, full of construction waste. She lifted it up over her head and tried to dump out its contents into the trash truck that was almost out of her reach. When she spilled out the bucket its contents missed the truck and fell back on her, spreading out on the ground. As I knelt down to help her pick up the trash, I first wanted to laugh and then I wanted to cry. She became so precious to me. I thought that some day she would be a grown woman and that she deserved every opportunity to live her life with dignity and become an honored member of God's family. I knew that I would always be her friend. I thanked God that I could help build her house, and I prayed that He would enable me to do more."

More Than a Habitat Person

Cynthia Lee was working as an AmeriCorps volunteer with Thibodaux Area Habitat for Humanity in Thibodaux, Louisiana, in 1997, when she met

Daphne Dominique. Cynthia tells a powerful story of the dramatic effect of friendship on that young woman.

"Daphne was a shy sixteen-year-old," Cynthia reported. "Her mother's home was due to begin construction in late March. Daphne was in the eighth grade, and the only thing she talked about was going offshore to be a cook when she turned eighteen.

"The Family Nurture chairperson asked me to take on Daphne. I knew this was going to be something I'd enjoy. I began by having her come into the office to answer phones and help out with filing. I began to witness some of God's miracles. Daphne quit school in March and by April she was enrolled in a GED program and feeling much more comfortable with teenagers in her peer group. I made myself more than a Habitat person to Daphne. I made myself a friend for her. As her family's house was being built, so was a young woman. I showed her a part of life she had never known. Before she and her family got involved with Habitat, Daphne only knew child rearing because her mother worked two jobs and Daphne had to take care of her little sister, Starla, most afternoons," she said.

"I was able to introduce Daphne to other things such as college life. I introduced her to some of my sorority sisters, the lovely ladies of Delta Sigma Theta. Eventually, Daphne took her pre-GED test and passed. And she has started a savings account with earnings from a part-time job. When Daphne's new Habitat house was dedicated," Cynthia said, "I was so happy to realize that the ministry had truly changed her life and also mine."

Before Christmas

Byron Higgin of Wild Rivers Habitat for Humanity in Siren, Wisconsin, told me a wonderful story of one very special friendship that had a profound effect on a new Habitat family. Dan and Carrie Hunter's old green trailer house was falling apart. In the winter, Dan had all he could do to scrape together enough money to pay the heating bill. The floor and ceiling were nearly gone, and they wondered how they would survive another winter.

"The kids have never had anyone over to play. They can't invite someone to a place like that," said Dan.

Then along came Willy, and their lives changed forever. Willy Holmberg is the building committee chairman for Wild Rivers Habitat in Burnett and Northern Polk Counties in Wisconsin.

Tears flowed down Carrie's cheeks when she heard that Wild Rivers

Habitat was going to build them a new home. But she had no idea of the kind of special relationships that would develop—or the changes that would take place in her four children.

It was September when Willy discovered the dilapidated old green trailer house. "I decided right then and there that I wasn't going to let that family stay another winter in that place," he said.

The retired construction firm owner dedicated himself to the task of putting the Hunters into their new home by Christmas. Each day Willy went to the site. He and the family were brought together closer as the weeks went by. "Hey, there's Grandpa Willy," one of the girls would shriek. "Willy! Willy! Willy!" went the chant as he walked in the door, followed by heart-melting hugs.

One day when Willy arrived at the new house, which at the time was nothing more than walls standing on concrete, he saw one of the girls inside. She had a broom in her hand and was smiling as she swept out her room, cleaning it up. Already the girls were proud of their new home, and it didn't even have a roof.

That night Dan and Carrie decided to take their family for a walk in the moonlight. They stood inside the home and dreamed about what it would be like to live there. When it was time to leave, the girls didn't want to go. "Can't we sleep here tonight?" they asked.

Into the Hunters' lives came some very special friends—all Habitat for Humanity friends who were as serious as Willy about putting the Hunters into a home by Christmas, as Byron explained it. Joanne Grefsrud became the Habitat representative to the family. "This is a family that realizes they're going to be able to invite friends to their home. They've started going to church, and they're getting very serious about their lives," she said.

"But Dan had a deep concern. 'We've gained so many friends and we enjoy them so much, but when the house is finished, we know they'll go away. That will be a sad day,' he said.

"The whole family cheered, though, when they found that Joanne wouldn't leave them—that she would stay close to them even after the house was completed, that she'd keep on visiting with the family

"'This relationship isn't for a few months or even a year. I know it's a life-long commitment,' Joanne said."

Just before Christmas, the house was finished. On the day the house was dedicated, Dan stood outside, lit up a cigarette, and said, "I promised the family I'd never smoke inside this house. Maybe I can quit altogether." He put his head in his hands and exclaimed, "Do you know what this means to us today? Remember your most favorite Christmas? Now multiply that by one million. That's how it feels."

He threw down the cigarette and stepped inside the door. Immediately he was hit by a happy, jubilant child who looked into her father's eyes and delivered a smile that went straight to Dan's soul. "Willy saw that smile, and he knew what it meant," wrote Byron. "He had lived up to his promise and now it was Christmas-present time. He was there to give this home, this Christmas present to the Hunter family.

"But it was Willy who was receiving," added Byron. "His heart was as warm as toast, his eyes moist, and his arms full of Hunter children. He felt so blessed. Not many eyes were dry that day. Carrie's weren't when she saw the almost-new living room couch on the front lawn. The kids' weren't as they jumped onto their new bunk beds, beady eyes shining brightly, smiling at visitors as they passed by the door. And neither were mine—as suddenly from deep down within my soul I discovered something I never expected. Here I was, a Habitat for Humanity board member who spent years bringing this first home to reality—and all the time I was fulfilling my own dream—to help someone else. It wasn't my own money, but the loving dollars of others I helped use to bring peace, happiness, and joy to one family.

"Since that day, teachers in the Hunter kids' school say there has been a dramatic change in the children. They're more outgoing, they get along better, they join in more, and their grades have improved," Byron ended. "And by the way, some little girls from school are spending the night at their house tonight."

Habitat Man

David Wadsworth is a longtime Habitat partner who started his work with this ministry when he was a pastor in Kingsport, Tennessee, and now serves as pastor of St. Giles Presbyterian Church in Greenville, South Carolina. David tells an exciting story of a little friend he met at the work site of a house that was sponsored by the church. "There was a lull in the construction work, so I began playing with six-year-old Keaun, one of the children of our partner family. Keaun loves superheroes and has a great imagination," said David. "That afternoon, Keaun had dressed himself up as a carpenter. He found a hammer, a tape measure, a nail apron, safety goggles, and a St. Giles commemorative carpenter's pencil. When a small group of other workers gathered around to watch Keaun and me hammering nails into boards that would become our pretend doghouse, he became a little shy and self-conscious. He motioned for me to lean over to listen to something he wanted to tell me.

"I leaned over and he whispered, 'Tell them who I am.'

"'Okay, Keaun, I will be glad to tell them who you are. But who are you?' I asked.

"A huge, priceless grin covered his whole face. With a sparkle in his eyes Keaun said, 'You know who I am, Preacher David. I'm Habitat Man!'

"'Well, it takes one to know one, my young friend!' I said. I hope and pray that Keaun gets to meet and work beside as many wonderful Habitat men, women, boys, and girls in his lifetime as I have in mine!"

A Boy's Window and Smile

Allison Williams, a Habitat volunteer from Athens, Ohio, encountered a nine-year-old boy on a work site in the nearby rural town of Amesville. Even though the encounter was relatively brief, a bond of friendship was made and the whole overall experience was life-changing for Allison.

"Orange, yellow, and red maple leaves sprinkled the surrounding hills," Allison recalled. "The autumn beauty of southeastern Ohio should have sparked my imagination, but on that day, I barely noticed it. All I could see was the innocent smile of a nine-year-old boy.

"The other volunteers and I drove for miles up the rocky, winding, dangerous road that led to the work site. I didn't know how badly the family needed new living quarters. I gazed out of the car window at the cramped, dilapidated trailer. Four people lived in three rooms without plumbing or electricity. Somewhere deep down I know people actually endured those conditions, but I had forgotten about this unfortunate reality until I witnessed it firsthand. My first instinct was to feel pity; I wanted to reach out and apologize for being blind to their obvious 'misfortune.'

"Construction was underway on a lot adjacent to their trailer. My first task was to help install windows. As I searched for a hammer, I noticed the big, brown eyes and jet-black hair of a little boy standing next to me. His hair was tousled from playing and working outside all day, and as most young boys, he somehow managed to get most of his body smudged with dirt. His faded jeans and gray sweatshirt were streaked with grass. As he reached up, he not only gave me a hammer, but also an eager, genuine smile.

"At that very moment, an indescribable feeling flooded my soul," explained Allison. "His smile captured the beauty of childhood and represented all that was good and pure in human nature. Psychologists say the smile of enjoyment is different from the smile of accomplishment or recognition, because it is unexpected and therefore completely genuine. And on that day, both of us were wearing the same smile, but for different reasons.

"His smile was sparked by the prospect of building his home. I smiled not only because I was helping his family, but because I finally understood what motivated me to do so. We were both contributing to something bigger than ourselves. Bringing us supplies or carrying small boards was an honor for him, and it was an honor for me to nail the boards and windows into his soon-to-be home. The smiles that resulted from these actions were a common bond between us.

"As I fit the window into its designated frame, I realized that the room on the other side was most likely going to be the boy's new bedroom," Allison continued. "He had probably never had his own room—a special place that he could decorate with toys and go to when he wanted to escape into a fantasy world. I envisioned winter mornings when he would jump out of bed at the crack of dawn and peer out his window at the fresh layer of snow blanketing the ground. I realized that sheet of glass was more than the dictionary definition, which is 'an opening constructed in a wall to admit light or air.' As Michael Pollan, editor-at-large of *Harper's* magazine, noted: 'Look at a window long enough, and you will begin to see that it is not only a material object, but the embodiment of a relationship—between the self on one side of the glass and the wider world on the other.' This was a window of opportunity; it symbolized this family's new beginning. With his family's worries somewhat lessened, the boy could now peer out of his window and appreciate the exquisite landscape of the surrounding nature. He now had a reason to believe that everything was possible, and he could smile a little more often.

"For me, installing the windows was the most meaningful task I could have ever been given. It provided the opportunity for me to look at life in a whole new light, to believe in the power of dreams and fresh beginnings. Looking through the clear glass allowed me to step into the boy's world; I could relate on a whole new level because I understood what this new house would mean for him and his family. This new perspective made me smile on the inside," she added. "It wasn't a smile that I would have routinely flashed for a snapshot, or the smile I tentatively gave when people wanted to see what my braces-free teeth finally looked like after three long years. It is a smile that returns often when I reflect on my enjoyable and meaningful experience. I hope the boy thinks of me and smiles too when he looks through his new bedroom window at the maple-covered Ohio hills."

Almost Heaven Runners

Another remarkable experience transpired just to the east of southeastern Ohio in Circleville, West Virginia. Almost Heaven Habitat for Humanity Executive Director, Kirk Lynum-Barner, has built one of the most remarkable affiliates in the country. Under his leadership, Almost Heaven Habitat for Humanity has built nearly a hundred houses for low-income families. And he has accomplished this remarkable feat largely with resources from outside the area. The county has a small, low-income population. So Kirk and his Habitat partners have developed an extensive program for bringing in work groups to help with the building. Inevitably, deep friendships result.

Michael J. Kasper of Middleburg, Ohio, tells a touching story about his Habitat experiences in West Virginia and of friendships that emerged there. "I was in Circleville with a work group from my church. Our assignment was to build a straw bale house, complete with electrical and plumbing facilities for a family of nine—the Van Scoys—who had been burned out of their house during the winter.

"My job was to haul sixty-pound buckets of stucco to the house site. At the end of each day, when our group reached the van, we all were exhausted and ready to nap or meditate. In the afternoons, after work, and in the evenings, we did sightseeing or enjoyed musical events. The night we arrived, I square-danced and clapped to the music of a folk band. Some evenings I ran through the countryside, reflecting on the work and people I had encountered that day.

"I believe that a big part of Habitat is about getting to know yourself," he realized. "This organization helped me to clarify my values and make me more appreciative of all the wonderful things I have. Habitat has given me a feeling of pride and self-worth."

It also gave him new friends. "David Van Scoy and his family became friends and family to our volunteers during the week," he said.

"When Habitat for Humanity people left after a six-hour workday, David was still working on his house until long after dark,"Michael recalled. "David, Michelle, and their children showed much enthusiasm and participated to see their dream come true. As a believer in the Christian faith, I enjoyed hearing from David about the growth in his faith to believe and praise God. David and I are both runners, and I find runners have a certain bond with each other. Because David has become one of my role models, I have decided to run at college. I was amazed that someone who was a stranger at the beginning of the week could feel like family by the end of the week. Saying good-bye was tough. The greatest gift in life is the opportunity to change somebody's life. My Habitat experience was truly life changing."

Overwhelming Effect

Sometimes the friends you meet and their effect on your work is so overwhelming, you just have to express it in broad terms. That was the case for Frank Stell Jr. of Freeburn, Kentucky. "I was among those who organized Phelps Area Habitat for Humanity. Our first project was putting a roof on a house for a family," he said. "That work gave me such a great feeling of knowing that I had helped someone in need. I have met a number of people who have become some of my closest friends, and they have made an impact on my life that I will cherish always.

"I have given a lot of hours, sweat, tears, and prayers to a cause that will always be close to my heart. A lot of people have asked me why I spend all that time working for no money. My answer was and will always be, 'When I hand the key to a new home to a family who never had a decent home before, then all the time I spent is worth more than all the money in the world to me. And as long as I have breath in my body, Habitat will be part of my life and I will continue to contribute in whatever way I can."

Anthony McMahon, a dedicated volunteer with Charlotte, North Carolina, Habitat for Humanity is another of those who could only write in broad, exciting terms about a Habitat experience he found both rewarding and filled with new friends. "Habitat gives back to its volunteers," he wrote. "It gives them an opportunity to learn new skills and then teach those skills to others. It gives you a tremendous feeling of self-worth. It gives you the chance to meet literally hundreds of new friends, and it teaches you that you don't have to make a large salary to enjoy your work. And most of all it makes you feel so happy."

A Fiftieth-Birthday Building Bash

Sometimes a Habitat house can be the reason old friends get back together. Mary Smith of Atlanta decided to do something very unique with her friends to celebrate her fiftieth birthday. In so doing, many new friendships emerged, and a lot of work got done on a Habitat house.

Mary decided to have her very special event at Beaufort, South Carolina in partnership with Low Country Habitat for Humanity. Twenty of her friends were invited, and they came from Pennsylvania, Wisconsin, Georgia, and other places. None of the women had any construction experience, but Low Country Habitat provided the leadership to assist the women in putting up framing and performing other construction tasks.

Some of the women had never met, since they came from different times in Mary's life. So many new friendships were formed. All of the women said that they thoroughly enjoyed the experience.

Below is a portion of the clever invitation Mary Smith sent out to invite her friends to the build in Beaufort:

I had this idea, as soon I turn fifty
To gather my friends and do something nifty.
But what could we do to suit everyone's needs
Should we party or travel, drink wine or read?
Take in a play or enjoy a brisk walk,
Bring up old memories, laugh and just talk?
Then it came to me finally—let's do it all!
And do some goodwill while we're having a ball.
I'll gather the troops mixing old friends and new—
Throw in family members to sweeten the stew.
All will get along fine, no reason to dread
'Cause they're all my best friends—they have that common thread.
So here is the plan—invitees only female,
Let me know soon via phone, fax, or e-mail.
From Saturday the second through the sixth day in May
You're invited to Beaufort, South Carolina, to stay.
We'll help build a house for people in need,
Hammer nails, paint a room, plant gardens with seeds.
No need to worry, you won't need a work skill
To see friends working together will be my great thrill.
The day is all set, so's the place and the time
And I'm sure you're quite sick of reading this rhyme.
Now I hope you can come (please let me know swiftly)
To celebrate with friends my birthday of fifty.

A Surprise of Old Friends

Sometimes Habitat is also a surprise vehicle for bringing old friends or acquaintances back together.

In Huntsville, Alabama, Tom Hilinski was participating in his first Habitat for Humanity blitz-build. He expected to meet some interesting people, but since he was not affiliated with any organization and since he was from the North, he did not expect to meet anyone that he knew working on the site. This is how Tom told his friendship story:

First, I discovered that two of my work companions were Yankees like me. One was born in the same hospital as I, and the other was born and raised in a small town not more than seventy miles from my homestead. However, these "small-world" events were just a warmup for the big surprise I had coming.

I was working outside an open window of the house when I heard a voice inside mention "University of Pennsylvania." At first, I was going to let it pass, but I decided to see if I needed to rescue the good name of my alma mater in the South.

"Who's talking about my alma mater in there?" I yelled.

The voice came back, "I graduated in 1963."

I yelled back, "I graduated in 1964 from the Moore School of Engineering."

After a brief pause, the guy appeared at the window and said, "I graduated from the Moore School too."

At a distance of ten feet, we both looked carefully at each other through the open window, wiped away thirty-four years of aging, and simultaneously recognized each other.

"Roy Adams!" I said as he said my name at the same time. We had not seen each other since 1963. We had worked together for about a year, with desks side by side, doing engineering and math work with electro-mechanical calculators in the early days of the computer age. Our relationship now brackets those years and puts nice bookends for me on my career as an engineer and for the experience of the blitz-build. We remembered a lot of memories about our youth just entering the field of engineering. Now we also have a lot of new stories to trade. We have Blitz-Build '97 and Habitat for Humanity of Madison County to thank for rejoining two old engineering colleagues.

A Thriftshop Full of Friends

Janet Harns of Jackson County Habitat for Humanity, headquartered in Marianna, Florida, made a network of friends from the creation of a thrift-shop for that affiliate. "And the story of the shop is a miracle," she said. "Many retired persons who felt that they had nothing to offer any longer came together and worked, establishing networks of friends and serving the community. More than fifty persons volunteer each week, working assigned hours. The shop earns enough funds now for nearly three houses per year. The effects of purposeful activity in a situation where they can see the results have been inspiring. Volunteers testify to the changes in their lives with a purpose to get up and go to work. They transport each other to doctors' appointments and check on each other. They socialize and attend churches together. All ages and races interact and work together to meet a goal. In fact, they set a sales goal each year and then strive to meet it. They have become a Habitat family."

Brought Me Closer

Some of the friendships actually happen between family members who learn a lot about each other as they build side by side with new friends. One of our very generous partners in the ministry of Habitat for Humanity is Willow Creek Church in Barrington, Illinois. That very large congregation has long been supportive of this work, both in Chicago and in the Caribbean. Since 1995, the church has been sending work teams to the Dominican Republic. On those annual trips, many friendships have formed. One year a reporter from the church was interviewing a homeowner, Cesar Feliz, for an article in the church newsletter. The reporter asked Cesar why he worked so hard at the site where several Habitat houses were being built. Cesar worked six days a week, even though the requirement was only two days a week. His answer was a wonderful surprise.

"There are two reasons," he began. "First, I have made friends, and that new community of friends will help me in the future because I am helping them now. Second, this work has brought me closer to my family. My wife didn't think I could do what I'm doing now. She thought she would be building this house alone, without me. Now she has more faith in me. She trusts me. She tells me that now I'm the man she wanted to marry."

Never Finished with Habitat

And many friendships grow and remain between volunteers and new home-owners. Love and friendship grew up between homeowner Pamela Ford and various Habitat volunteers while her house was being built. Theresa Wise, executive director of Habitat for Humanity of the Greater Erie area in Pennsylvania shared that Pam's experiences were deeply meaningful to her and to all the volunteers. Pam expressed it best in her own words to the new friends she made:

> I walk so proud these days because of what you taught me. I never thought people could care so much about a person they never knew, but now there is no doubt, just love and tears. I brag so much about Habitat to people. Thanks, Diane and Mary. Thanks Nancy, John, Don, Frank, Gordon, Herman, Terry, and all the rest.
>
> And hey, Margaret, you thought I forgot about you. I could never forget a woman who made me feel like her own family, a woman who has seen my sad times and my happy times. It's hard for me to tell my feelings to folks, but you are my family, you are my God-sent family. I love you with all my heart. I'm not finished with Habitat, not now or ever. If you ever need me for anything, please call.

A Friend to All

Jenny Borland of Habitat for Humanity of the Gallatin Valley, Bozeman, Montana, told a moving story about a volunteer who became a friend to all with whom he worked. "Bob Murray retired from his job early and became bored. He called the Habitat office to see if we had a job for him, and we did," Jenny said. "He started volunteering and liked it. He came every week for two years. Bob was a quiet guy with a wonderful dry sense of humor that made it fun for everyone to be on site with him. After working for about a year, he decided we needed some ladders, large saws, and other items. So he wrote a grant application to IBM for a thousand dollars to purchase these needed tools. When the money came, he and the construction supervisor, Earl Goldsworthy, were like little kids, seeing how much they could get with the money.

"One day, though," Jenny recalled, "Bob became our first volunteer to be injured. Three of his fingers had been cut off except for a thin strip of skin. I was so upset. I ran to the hospital and then to find his wife. When I found her, she was the calm one, comforting me and assuring me that Bob would be fine—and he was—after surgery and two months off from working with us. They were able to reattach the fingers, and by the time he returned, his fingers were only a little stiff. We were so glad to have him back.

"Then in the fall of 1996," Jenny went on, "he was telling everyone on site how he was going fishing one more time before putting up the boat for the winter. The next day he was dead. He passed away quietly in his chair watching television. Habitat families living in houses he had helped to build, building committee members, the board, and other volunteers who had built with him were all in shock. Everyone came to his wake to tell funny stories, to share appreciation for this wonderful friend, and to comfort his family. Bob Murray was devoted to Habitat and freely gave of his time, talent, and humor. He touched us all and truly embodied the Habitat spirit."

A Circle of Friends

In south Florida, the small but very dynamic affiliate of Boca-Delray has built a total of more than thirty homes in Boca Raton, Delray Beach, and Boynton Beach. In 1996, they initiated a very exciting new program called Circle of Friends. This is a group of local church congregations that has come together in partnership with the affiliate to provide funds and volunteers to build houses. As of mid-1999, this unique Circle has built eight houses. And the Circle surely has brought together many old friends and has made a lot more new ones.

A High School Ring

In New Smyrna Beach, Florida, Lee Hoover has long been a wonderfully dedicated partner and leader of the local affiliate, Habitat for Humanity of Southeast Volusia. She attended a regional Habitat conference in Valdosta, Georgia, a few years ago. There she met Alberta Moore, who for years was our beloved receptionist at Habitat International headquarters. When Lee found out that Alberta was from Americus, she asked if she knew a certain man to whom she had been "sort of engaged" while she was in college.

"Yes," Alberta replied. "His sister is my best friend." Alberta enabled Lee to make contact with her old beau. Lee wrote to him and asked if he still had her high school ring. She wanted it because she was going to a reunion in Baltimore. Two days later, Lee received the ring by registered mail in a box and in perfect shape.

Lee continued, "I wore the ring to the reunion, told the story, and everyone loved it. I talked to him later. He returned a good photo also and promised to keep in touch. He had had that ring for over fifty years. How about Habitat for Humanity getting two old friends together! We are both too old for our mates to be jealous—both happily married for more than fifty years, but the reunion was such fun because of the ring."

Rosalia Thomas, Cathy Belatti, Phyllis Tutor, and Linda Fuller working hard on the All-Women's house at the 1998 Jimmy Carter work project in Houston, Texas. photo by Robert Baker.

Why We Do It—Remembering in Houston

The final story in this chapter is a powerful one from Paul Leonard, Habitat International board member of Mooresville, North Carolina. Paul and his wife, Judy, sponsored one of the houses at the Jimmy Carter One-Hundred-House Blitz-Build in Houston in June 1998. Paul recruited twenty of his friends to build the house. It was quite an experience for all of them.

Paul wrote an account of the week and titled it "I Remember Houston."

Following are excerpts from his remembering and a good snapshot of what it's like to build a house and make lifelong friends in the process:

The goals were clear: Frame and dry-in the house on Monday. Dry-wall, roof, and side the house on Tuesday. Paint and complete siding and roofing on Wednesday. Install interior trim, doors, and cabinets on Thursday. Landscape and dedicate on Friday. A piece of cake!

We cheated and started early. On Saturday and Sunday our crew leaders measured the prebuilt wall panels and drilled holes for the foundation bolts. We also sweated and got a foretaste of the week ahead as temperatures became oppressive by 10 A.M. By Sunday noon, the size and scope of our challenge became apparent. Building a house in five days was one challenge—building it in five days in 100-degree weather was another.

MONDAY

Monday wounded our pride. President Carter had said on Sunday night that no one would leave the site on Monday until all houses were dried in. When we left the site at 6:30 P.M., our house was not dried in. The trusses were in place but without roof sheathing and felt. Our minds and hearts were willing but our bodies would not move.

TUESDAY

Tuesday was a different story. The youngsters jumped on the roof, installed plywood and felt by 11 A.M. Now we were only four hours behind schedule. But we were running out of gas. An "angel" appeared and asked where we needed help. "Roofers," I replied. Within twenty minutes, a six-man skilled crew of roofers was on our roof, and within three hours, the roof was completed. Meanwhile, siding was underway on all four sides of the house. The plumbing and electrical subs had completed their "rough ins" and passed inspections. This was the good news. The bad news was that the HVAC subcontractor was nowhere in sight. Work on the interior stopped. No HVAC. [Heating, Ventilation, and Air Conditioning.] No insulation. No dry wall.

Another "angel" appeared. Our problems were not unique. The HVAC subs had pulled off the job Monday night because the houses were not ready and no one knew when they might return. The project leaders went on radio and television seeking HVAC subs. The entire project was threatened. Charles Simms with Beltway Mechanical walked up. Charles had heard the plea and had come to offer his help. Two crews were on the way. By 5 P.M. we called the inspector and began insulating the walls. When we left Tuesday, the house was ready for dry wall. Siding was 70 percent complete and roofing was 100 percent complete. Pride was back. We moved that house! However, the schedule called for dry wall to be finished. Our dry wall had not begun.

WEDNESDAY

On Wednesday, while most of the crew worked to complete the siding and soffit, a few of us started stocking the dry wall. Our dry-wall crew leader was getting organized when professional dry-wall hangers came to the front door. Someone has said that luck happens when preparation and opportunity meet. Thanks to the steady pace of our crew the house was ready for dry wall when the dry-wallers needed a place to start. Coincidence or house angel at work? We adopted the roofers, dry-wallers, plumbers, electricians, HVAC installers, everyone. We gave them water, shirts, and materials. They accepted us and checked on us even after completing their work.

All day Wednesday the crew worked outside and finished the siding. By now many of the crew had worked with Linda, our homeowner. She has four children and supports herself and them with her job as a nursing assistant at a local hospital. She was using a week's vacation to work with us and see this house come together. Jimmy Carter said that most of the homeowners probably did not believe on Sunday night that their home would be completed on Friday. I believe this was true of Linda. She didn't really smile until Wednesday. That's when she began to believe that this might really happen. She smiled when she told me that she liked the siding color. She smiled when we asked her opinion about attic stairs and kitchen cabinets. Slowly we were getting over the social and racial walls that separated us.

Wednesday was hump day. It was also the day that Harry Smith of CBS News visited the Sakowitz site. Harry visited our house that day. He wanted to know why we had come from North Carolina to build houses in 100-degree heat in Houston. He could not understand why twenty people would give up their vacations and pay their own way to Houston to build a house. I told him we all had "infectious habititus," that we liked working together, that we received more than we gave. I'm not sure he believed any of this. One of our volunteers, Kevin Strawn, heard Harry's question. "Why are you doing this?" Kevin's short answer was, "I would do anything for my friend Paul." Kevin reminded us that we were all friends, and that the standard Jesus set for us was to do anything for our friends. Kevin hoped that this was his motivation for going to Houston, to follow in the steps of the one who befriended all mankind and called us to be friends to each other, and if needed, to lay down our lives for our friends. I wish I could have made it that clear to Harry Smith.

THURSDAY

Thursday was paint day. In four hours we applied two coats of paint on walls and ceilings. Some of our crew went to help with other houses. Late Thursday we got back in the house and started hanging doors and installing base.

Friday was complete chaos. Part of the crew worked outside to plant trees and shrubs and lay sod. The rest of us worked inside to finish trim, hang kitchen and bath cabinets, install tops, and touch up paint. Plumbers, electricians, and HVAC mechanics were trimming out their work.

FRIDAY

Friday was dedication day and also the day we met Sammy, Terence, Brandon, and Niketa. These were Linda's children. The day felt like Christmas in June. It was Christmas for me when ten-year-old Brandon sang "Moving on Up," the theme song from the television program *The Jeffersons*. The best line was, "We finally got a piece of the pie." It was Christmas in June for me when Linda thanked us and told us how nervous and frightened she was on Monday. "You showed me how to do stuff without talking down to me," she said. "You made me feel important." And it was Christmas for me when seven-year-old Niketa came up beside me and looked into the window and exclaimed, with face beaming, "That's my room! May I go in and work? I want to help."

Words cannot describe those moments. I just know that all the hard work and sweat seem so insignificant beside the hope and happiness that dance in the eyes of those children.

Harry Smith, that's why we do it!

CHAPTER 10

Nurturing Faith

When I first came out to work here, I knew religion.
But now I know love.
—Habitat volunteer and homeowner

Habitat for Humanity has always been and continues to be openly and unashamedly a Christian ministry while remaining just as steadfastly open to all who want to participate in this faith-and-love venture. Homeowners are chosen without regard to race or religion, and volunteers and donors are invited to be a part of the work regardless of their faith commitment.

In every way possible, though, without pressure or coercion, God's love, the original impetus for Habitat's existence, is expressed. For example, at Habitat for Humanity International headquarters in Americus, Georgia, we start every workday with devotions, and attendance by staff and volunteers is encouraged but never required. This practice has been followed since the inception of the ministry more than two decades ago.

Devotions are also held at the beginning of each workday at the annual Jimmy Carter work projects. Then, at the end of the week, each house is dedicated and a Bible presented to the family. Those dedication services are followed by a big "Habitation" program, which is a profoundly religious event.

Generally, all local Habitat affiliates worldwide hold short devotionals prior to workdays, and upon completion of new or renovated houses, dedication services are held and Bibles are presented to the families.

Building on Faith

In September of each year, we hold our annual Building on Faith week. This event emphasizes the partnership between churches and other religious organizations and Habitat for Humanity. Typically, churches in various communities sponsor one or more Habitat houses to be built during this special week, which culminates on the third Sunday in September when Habitat's annual Day of Prayer and Action for Human Habitat is observed. The Day of Prayer and Action has been observed since 1983. That observance includes our emphasis in participating churches' sermons, "moments for missions" about poverty housing and Habitat's response to the problem, and bulletin inserts with litanies or responsive readings concerning substandard housing and homelessness. In 1998, two million bulletin inserts were sent out to an estimated twenty thousand churches.

Building on Faith week has been observed since 1995. The first year, ten affiliates participated. In 1996, the number of participating affiliates increased to seventy. In 1997 the number jumped to one hundred and ten, and in 1998 an estimated two hundred affiliates joined in observing this special week. The goal of our church relations department is to have at least five hundred affiliates participating in Building on Faith week in 1999. In the year 2000 we want to have *every* affiliate worldwide be a part of the special week and involve tens of thousands of churches to build a total of ten thousand houses.

Since the inception of Building on Faith week, I have visited several of the affiliates each year. Typically, I go to seven to ten of the affiliates that are a part of the special blitz-build week. I have always been struck by the strong spiritual emphasis at each of the work sites.

Habitat for Humanity International and various state, regional, and national Habitat for Humanity organizations are constantly holding events such as regional, state, area, or national meetings of many kinds. Invariably, these events have a strong Christian tone. Sessions begin with prayer, readings from the Bible, and a message about God's love.

I speak at many of those meetings all around the world. I also speak at many gatherings sponsored by other organizations, both Christian and secular. I always present Habitat for Humanity as a Christian movement, an incarnation of God's love.

Local Habitat affiliates also have events of many kinds. Again, the Christian roots and motivation are exposed and expounded.

Finally, individuals who love God and who are involved in this work out of their spiritual motivation interact with people who are at different levels of faith commitment at work sites, in meetings, and at other places. This personal sharing, again always free of pressure or coercion, often results in people deciding to change their lives spiritually and embrace the Lord.

To Love One Another

To me, all of the above, including the driving of the nails and the sawing of boards to actually build the houses, are a part of fulfilling the mandate of Christ to love one another. That is the essence of the Christian gospel. Love. Love one another. Love your neighbor as you love yourself. Invite strangers in. Share your resources. Give of your talents and abilities.

I believe that salvation comes from God. Salvation does not come from Habitat for Humanity, Millard Fuller, or any other individual or earthly organization. Salvation comes from God. Therefore, it is not incumbent upon me to save anybody. My duty as a Christian is to love, and that love must be universal. I must love people who are like me and those who are vastly different from me. I must love those who agree with me politically and theologically and those who don't. I must love my fellow Americans and people of other countries. I must love even my enemies.

This understanding of the universality of God's love and of our responsibility to love without limits is why, for example, we don't discriminate in selecting families to receive houses. We build for Christians, Muslims, Hindus, Jews, and for people of other faiths or no faith. We also build around the world and ask every local affiliate, through tithing, to help build in other countries as they build locally.

While I do not feel called to "save" people in a spiritual sense, I do think that it is right and even necessary to share our faith with others. And I believe that should be done with both word and deed. After all, the Bible says that we should "proclaim the gospel." As we do that, seeds sometimes fall on fertile ground in the hearts of people. The spirit of God begins to stir in their lives and foster change. To me, that is a joyous thing to see.

All of us are but broken vessels, full of fault and sin. But God uses imperfect people to accomplish His perfect will. This realization fills me with great hope and optimism. I am free from the burden of feeling that I must be perfect (even though the Bible says all of us should strive for that), and I am free of the onerous task of having to "save people." Instead, I can love with enthusiasm and with abandon. God is in control. Because I am a vessel in His hands, I can do my work knowing that the results are up to God.

My spiritual mentor, Clarence Jordan, gave me an insight about the call of God to love. He enlightened me to the truth that salvation comes from God and not from anyone else. That simple revelation has helped me spiritually more than anything else, because I was raised in a spiritual environment in Alabama that led me to believe that there was a great *burden on me* to save those who did not go to church (and it was put like that, since church membership was synonymous with salvation). I now believe fervently that I should love as best I can, without limits, and trust God to touch the hearts of those who are rejecting Him. God's love is without limits, and that means that it extends beyond the grave. That too is a very liberating spiritual insight from Clarence Jordan.

Millard Fuller working with Ohio governor George Voinovich on Habitat for Humanity International's seventy thousandth house during Building on Faith week in September 1998.

New Frontier in Christian Missions

I have always seen Habitat for Humanity as a new frontier in Christian missions, a creative and new way to proclaim the gospel and share God's love. We are not a church, but we *are* servants of the church—servants of the whole church, liberal and conservative, Catholic and Protestant. And we are a vehicle, a common meeting ground, where all people—Christian and Jew, Hindu and Muslim, Buddhist and Jain, faith-filled and spiritually empty—can come openly and joyfully to build together and have fellowship as we strive to eliminate poverty housing from the face of the earth. I get excited just writing about it. And my excitement heightens when I reflect on what is happening on this ever-expanding common meeting ground. The lives of people are being transformed—homeowners, volunteers, staff people, and others. And more than all the other essential ways that Habitat meets needs and changes lives, these transformations are at the very heart of the "more-than-houses" dynamic of Habitat for Humanity.

Your heart will be warmed and filled as you read the following incredible stories of lives being changed through the ministry of Habitat for Humanity.

A New Set of Values

Shawn is executive director of Mountaineer Habitat for Humanity in Charleston, West Virginia. In 1989, Shawn had a job selling commercial building products which, he said, gave him little or no satisfaction. He was married and had a car and a mortgage. His father had died two years earlier, and the family construction business died with him. By his own admission, Shawn was without direction in his life.

The birth of Shawn's first child gave him a new outlook. He wanted to find a way to make her life different. He felt that she needed a new set of values, not the same as those which he exhibited.

"I was not a church person, even though I had come from a very diverse religious background," Shawn said. "As a child, I had attended Baptist churches, Nazarene churches, and Churches of God. I had studied Jehovah's Witnesses, and I had communed with Presbyterians. I wasn't sure how to give my daughter a religious upbringing. Even though I always thought of myself as a Christian, I was convinced that there were no Christians who believed the way I did." Then Shawn encountered Habitat for Humanity.

"My first experience with Habitat was at a fund-raising committee meeting," he explained. "That night I met people who would change my life forever. I met people who were concerned with things other than themselves. I met people who seemed to believe that the concepts and ideas that Jesus taught had real meaning for their lives. These people were living out their faith on Tuesday night, not just Sunday morning! These folks were the kind of people I wanted to be like. I wanted to be with them, to work with them, and to get to know more about them.

"Before I knew it, I had taken an entire Saturday off and worked on the home of Ronald Payne. I worked side by side with him, installing insulation in the walls of his new Habitat house. When the day was over, Ronald thanked me for helping. I said, 'It's my pleasure,' and I was surprised to find that I really meant it! It was one of the most pleasurable days I had ever had. It was probably the first day in my life that I had ever spent totally helping someone else," Shawn explained. "I soon began working as much on Habitat-related things as I did my sales job. The more I involved myself, the more I wanted to be involved. I was asked to serve on the board of directors, and I was honored to do so. I was elected as vice president and then president."

In the meantime, his spiritual longings had caused him to find a church home. He joined First Christian Church (Disciples of Christ) and was baptized in June 1990. "Without Habitat, I never would have made this commitment," he said. "Working for the church and Habitat gave my life new direction. Around the time when my second child was born in 1992, I prayed that God would show me how I could rearrange my life to devote more time to helping people through Habitat. I thought I was just asking God to rearrange my schedule, but it turned out that God had bigger plans.

"First of all, my commitment to the church led me to enter the ministry. A spiritual mentor suggested that by this I could help struggling smaller churches as a bivocational pastor. I was eager to do whatever I could to help the church. So in 1994, after going through all the ecclesiastical procedures, I was licensed to practice ministry in the Christian Church (Disciples of Christ) in West Virginia. After serving as a supply preacher at several congregations, I knew that this was what God wanted me to do. But I soon recognized that my job would always limit me in doing God's work. It was then that God played the trump card."

As president of the Habitat Board, it was natural that the current executive director, Bill Londeree, would come to Shawn to give notice that he would be relocating to another state, and therefore would be leaving Habitat. "Bill was one of the founders of our local affiliate, and for many people, including me, he *was* Habitat. For him to leave was something the board had never imagined," Shawn said. "Just when I was reeling from the shock of his announcement, Bill said, 'I think you would be the perfect person to take over this job.'"

Shawn was very skeptical about filling the shoes of this man he admired so much. "After praying a lot, I decided that this was God answering my prayer to have more time to devote to Habitat. I had faith that God wouldn't lead me into a situation I couldn't handle. I accepted the job. It is now five years later, and I have the best job in the world," he said. "It is full of hard, long days and emotional ups and downs. I have laughed and cried tears of joy and pain. I have never been happier. The flexibility of Habitat allows me to serve the church as an associate minister, and the things I have learned through Habitat, such as the writings of Clarence Jordan, have served to strengthen my ministry. I can only imagine what God has in store next. Habitat has led me into a life of serving others, into a personal relationship with Christ and into a career that is more rewarding than any I could have ever invented for myself."

One in a Million

Sometimes a dramatic "God incident" coupled with meaningful work with Habitat for Humanity brings a person to God. That's what happened to Craig Kelley of Knoxville, Iowa.

In the summer of 1996, Craig joined a work group from First United Methodist Church in Knoxville to spend a week in Chicago working with Uptown Habitat for Humanity. Craig's wife had been attending First Methodist. Craig was interested in the project because he had been serving on a committee for the city of Knoxville to study housing needs. He learned from his service on the committee that there was a need for housing for low-income families in Knoxville. He wanted to see how Habitat for Humanity worked in another location with the idea of possibly starting an affiliate in his hometown.

Fourteen people from First Methodist went to Chicago on July 7. They drove up in Craig's van. Upon arrival, the group was housed in a Catholic church in a very dilapidated neighborhood. Craig said that there were stripped cars on the streets and razor wire on top of fences.

The work was hanging dry wall in a three-story dwelling that was being rehabilitated. The group found the work to be fun and rewarding.

On Tuesday, Craig went outside the building to get a first-aid kit from his van and was shocked to see . . . nothing. The only thing remaining of his van was a pile of glass in the gutter.

Craig rushed back inside, telling everyone that the van had apparently been stolen. At that moment, though, the doorbell rang. A Chicago police officer was at the door looking for Craig.

The officer had recovered the van. Only fourteen miles had been added to the odometer. Two juvenile girls had stolen the van for a gang initiation. On the way to the station to pick up the van, the officer told Craig that the chance of his getting his van back with so little damage was one in a million.

The theft and quick return of the van, added to everything else that happened in Chicago working with Habitat, convinced Craig that God was speaking to him.

"After returning home," Craig said, "my wife and I took the needed steps to become members of the church. Up to this point in my life I hadn't been active in church since my teenage years."

Craig nurtured interest back in Knoxville in Habitat for Humanity, and his efforts worked. On August 1, 1997, Habitat for Humanity of Marion County, Iowa, became an official affiliate of Habitat International.

There Was a God

The Loudon County, Tennessee, Habitat affiliate has had several wonderful experiences with homeowners coming to the Lord through the Habitat ministry. Karol Massingill, a single mother with three children, was chosen to receive the first house built by Loudon County Habitat. Her old house had been severely damaged in a tornado.

Local churches and several from out of town responded to appeals for help from Habitat leader Hugh Brashear. Kingsport Baptist Men came with a good crew. And the house seemed to just jump up as local volunteers joined them.

One of the local volunteers was Bo Lewis. He worked on Monday and Tuesday, and the Baptist crew fell in love with him. They were surprised, though, when he showed up on Wednesday dressed in his garb as an Episcopal priest.

At lunch that day, the Baptist crew declared that Bo was okay. There was a lot of bantering as the men worked and fellowshipped together.

Hugh asked the men how they liked lunch. It was great, they exclaimed. "Well," he replied, "thanks are due to the Catholic church."

"How was lunch yesterday?" he inquired. Extra good, they agreed. "The Pentecostal church prepared and served you."

Hugh presented certificates of appreciation to all members of the crew at the end of the week in a moving program. Karol had come to love each crew member like members of her family. Her tears of joy and gratitude prompted many tears from the men. As the volunteers from Kingsport left, they told Hugh, "We got far more from this week of work than anyone here plus we came to know what is really meant by the body of Christ."

In the ensuing days, more volunteers came to continue the work of Karol's house. It was finished and dedicated thirty-seven days after beginning the framing.

One evening just before the work on the house was completed, a group of volunteers and Hugh Brashear were in Karol's new kitchen having a bite to eat. Karol came up to Hugh and exclaimed, "I had come to believe there was no God, but obviously I was wrong. Where should I go to church?"

Hugh replied, "Go home tonight and ask God where He would have you go."

She responded, "You're kidding, aren't you?"

"Give it a try," he said.

The next day Karol got back in touch with Hugh and reported, "God didn't say anything, but I keep thinking about Calvary Baptist, a couple of blocks from here. Do you know anything about that church? Is that where I'm supposed to go?"

"Take your family and go next Sunday. Go to the Sunday School and regular worship service," Hugh advised.

Karol did as she was told and later exclaimed, "It was perfect for the whole family. My children loved their classes, and I was particularly happy with the worship service."

A couple of weeks later the entire family was seen wearing "Jesus" T-shirts!

A Skeptical Man

The affiliate's family selection committee had a difficult time deciding on the Green family. Frank, the father, had a bad reputation as a mean man who couldn't get along with anyone. But the application clearly showed that the family met the qualifications to be a Habitat partner family.

A couple of people were given the task of going out to visit with the family. That visit revealed that the family was living in miserable conditions. The visit also confirmed what they had heard. Frank was a tough man. He had lost a leg and apparently had a generally rough time of it along the way. He was skeptical and cynical about everything. He said nobody had ever done anything for him, and he doubted they ever would.

The family did own a plot of land out in the county. After much discussion, Frank agreed to have Habitat build them a house there. About that time, Frank was diagnosed with lung cancer.

As the new house got underway, Frank was still skeptical about the motives of the Habitat people. He seemed intrigued by all the people who were coming out to work, especially the twenty college students who came from South Carolina on their spring break to work for a week.

One day Frank decided that the whole thing was a scheme to steal his land. He went to a local lawyer who, amazingly, agreed with him. Frank became almost impossible to deal with. He did all he could to run the volunteers off. Hugh Brashear said it was so tempting to just throw up their hands and quit, but they didn't because they believed they were building for the Lord.

Frank continued to protest as the building continued. Some of the Habitat people tried to talk to him about Christ. He wasn't interested.

On the day the completed house was dedicated, a banner was unfurled at the site proclaiming that the house was an expression of God's love. Frank said he understood but still resisted openly embracing God.

The family moved into their new home. Frank seemed to have a change of attitude. He told visitors about the "good people of Habitat" who built their beautiful home. He loved to show people all through the house.

Six months after moving in, Frank died. At his funeral, a local Baptist preacher officiated. He told how he and Frank used to have almost weekly conversations about his salvation when he visited Frank, but Frank continued to reject God. The pastor then related that on the morning of Frank's death, Frank woke up, looked into his wife's face, gave his heart to Jesus, and declared that he was going to a better place.

The Real Meaning of Life

Tina and Jason Pranger and their five children lived in a derelict two-room house. Jason, who worked at a fast-food place, did not make enough money to get a bank loan for a house. The family clearly qualified for a Habitat house. They were chosen.

The house was built in a week in June 1997 as a part of the Echo blitz-build in connection with the Hammering in the Hills Jimmy Carter work project in Tennessee and Kentucky.

As the house was literally springing up from the earth that week, the Prangers realized their lives were changing for the better. They came to realize that God was there along with the dedicated volunteers and that He was calling them into a personal relationship with Him.

Jason gave his heart to the Lord Jesus that week. "I'm saved," he declared to Hugh. "I now know the real meaning of life. I can now love my wife and family as I never could before. The church home we found is just perfect."

In the Midst of Building

Other homeowners have been converted in the process of building their Habitat house. In the summer of 1998, I was at the Jimmy Carter work project in Houston, Texas, where six thousand volunteer builders were putting up a hundred houses in a week. Toward the end of the week, I was making the rounds of some of the houses. The CEO of one of the sponsoring corporations came running up to me to exclaim that their homeowner, a single mother, had just announced that she was giving her life to God, that she wanted to follow Jesus.

In Chicago, the Uptown affiliate renovated a house for a poor woman and her aged mother. The family was too poor to move out during the renovation work so they just shifted the furniture and other belongings from room to room as the work was done.

Over a period of a few months the work was accomplished. During that time, everything in the house was a mess. Noise was constant during the workdays, and dust was flying everywhere. Finally, the job was nearly complete. Jim Lundeen, the affiliate director, went to the woman and told her they would soon be finished with the work. He was sure she would be glad to get their lives back to normal.

"No," she replied. She was not happy that the work would soon be over.

"Why?" Jim inquired. He didn't understand.

"I'll miss all of you," she said. "I've loved having all of these wonderful people here doing this renovation work. But I don't understand one thing. Why did you people do this for us? We are not Christians. Why did you Christians want to help us?"

Jim explained that God's love extends to everyone. "God's love has no limits," he said. "God loves you."

The woman began to cry. She was deeply touched.

A few days later the woman talked to Jim again, questioning him in more detail about the work, about why people had been willing to help them even though they were not Christians. Jim patiently explained that the motivation was love and that he and the others wanted to be used by God to express His love to them. He reiterated that God's love knows no bounds and that God loved her.

She wept again. Amidst her tears, she told Jim she wanted to be a

Christian and asked him to put her in touch with a church so she could be baptized.

Jim explained that they had not done the work to convert her. They simply desired to show God's love to her and to her mother. But *she wanted* to be baptized. She wanted her life to be characterized by the kind of self-less love for others she had experienced from the Habitat volunteers. Now the changed and happy homeowner is an active member of a church in her neighborhood.

At Habitat Headquarters

A classic hymn that is often sung at devotions in Americus at Habitat International headquarters is "I Have Decided to Follow Jesus." Over the years, several people have made decisions to follow Jesus while serving at Habitat headquarters. One such person was a student. She worked as a volunteer at our Educare Center and on the construction crew. After several months of service there, she returned to her home in Buffalo, New York.

Linda and I went to Buffalo for some speaking engagements not long after her return there. She met us at the airport.

Excitedly, she exclaimed, "I am now a Christian!"

"That's wonderful," I responded. "How did it happen?"

"Walter Money led me to Christ."

Walter is one of our outstanding construction leaders in Americus. Quietly and without fanfare, he had talked to her about the Lord, and she was convinced that she wanted to follow Christ in her life. It was that simple and that profound.

Another volunteer had come to Americus as an atheist. His wife was a dedicated Christian, but he didn't believe God existed.

At one point in his volunteer service, this young man was having a difficult time with another person on staff. A friend quoted Paul to him, from the New Testament: "'My grace is sufficient for you, for my power is made perfect in weakness.' Therefore, I will boast all the more gladly about my weaknesses, so that Christ's power may rest on me. That is why, for Christ's sake, I delight in weaknesses, in insults, in hardships, in persecutions, in difficulties. For when I am weak, then I am strong" (2 Cor. 12:9–10 NIV).

Around that time, the young man began to say that he was an agnostic and not an atheist. He would sometimes go to hear another volunteer preach at the local Mennonite Church. He began to say that he'd like "this Christianity thing if it weren't for the God part."

He started an e-mail discussion with one of our area directors about agnosticism. This director shared his understanding of faith and belief in God. The director recommended a book, which he bought and read. The director helped the young man to understand that it was okay to question, to feel challenged by what Scripture said, and that believing in God did not require believing every single word in the Bible.

A bit later on, some confusion at work stirred a lot of emotions in the doubting volunteer. It created many conversations with several fellow volunteers and staff people. That was in late 1997. Somewhere in the midst of all of that, as his wife reported, he found God—or put more correctly perhaps, God found him.

He and his wife left Americus to work elsewhere, but he returned to participate in a Holy Week blitz-build in Americus in March 1998. Twenty houses were put up in a week by eleven hundred volunteers. The now-believing man didn't stay until the end of the week, though, because he needed to return to his hometown to be baptized and receive holy communion at the Easter vigil mass at his Catholic church.

In Poland

Ewa Sikora, staff worker with the Christian Students Association, reported on a work camp of Polish and American students who built and fellowshipped together at the Habitat for Humanity affiliate in Gliwice, in southern Poland. She told of a young Polish volunteer named Siergiej who came to God in the course of that experience.

"Full of sacrifice and enthusiasm, the work of Christian American students, who labored together with unbelieving Polish students, strengthened the witness of a living faith by praising God through singing and working for other people," she said. "Siergiej, who had joined our Polish-American group in a spontaneous way, made his final decision to follow Christ in his life while at the Gliwice site."

A Prisoner Sentenced to Habitat

In Australia, a young prisoner came to Christ through his work with Habitat for Humanity. The Mornington Peninsula Habitat affiliate has been accredited as a charitable organization with the community correctional services in the state of Victoria, and Habitat had begun work with some of the convicts sent their way.

"Daniel, a young man in his early twenties, was convicted of larceny and sentenced by the court to one hundred hours of unpaid community service with Habitat for Humanity," explained the affiliate's Howard Fearn-Wannan. "Christians among the volunteers took an interest in Daniel, and without any overt preaching, witnessed quietly to him about their faith in Christ by deeds of self-sacrifice and acts of kindness to strangers. One day Daniel said to the construction supervisor, 'I've really enjoyed working with you guys on this Habitat project. I'm really impressed. You seem to enjoy this.'

Millard Fuller presents a Bible to new Habitat homeowners in Chisumbu, Zambia, in January 1998. Photo by Linda Fuller.

"When we explained that it was trust in Jesus and a preparedness to do what He commands that gives peace and purpose in our lives, Daniel responded, 'I want to follow Christ too. What do I have to do?'

"Daniel gave his heart to Jesus as his Savior and Lord," said Fearn-Wannan, "and he has become a new man with a bright future."

Habitat Changed My Life and Values

I met Yukari Mizuno in Japan in July 1998 when Asia director Steve Weir, my wife, Linda, my daughter Faith, and I visited Habitat for Humanity campus chapters there at several universities. I learned at that time that

Yukari had become a Christian as a result of her work with Habitat for Humanity. I also learned that she was going to the Philippines to be a part of the team that would prepare for the Jimmy Carter blitz-build in that country in March 1999. I asked Yukari to write to me about her experiences with Habitat for Humanity and about her decision to become a Christian. Following are excerpts of her letter to me.

"The reasons I decided to be a Christian were my experiences in the Philippines and your book, *A Simple, Decent Place to Live,*" she wrote. "I felt many spiritual things through short-term mission, but it all felt unclear. However, when I read that book, I was impressed and thought this is what I want to know and be involved in. I decided to be a Christian very naturally. Habitat changed my life and values totally. The dimension of things I got through short-term mission and my present job is immeasurable."

God Coming to Her

Our dear friend Fran Collier tells a wonderful story, not of her coming to God, but of the Lord coming to her. Fran was the driving force in forming one of the early Habitat affiliates in the United States, Habitat for Humanity of Greater Memphis, Tennessee. Later she and her husband, Wade, moved to Key Largo, Florida, and became very actively involved with Habitat there and most recently in Sarasota, Florida.

Early in the development of the Memphis affiliate, the affiliate made plans to build Habitat homes in Rossville, some thirty miles east of Memphis. When these plans were announced, white land owners in the area strongly objected. One woman who had property for sale refused to sell to Habitat because she had heard that blacks would live in the houses. She exclaimed, "What would my neighbors think if I sold to colored people?"

Fran says she yearned to ask her if she wasn't concerned more with what God would think. Fran said she was deeply distressed and discouraged.

"That night as I sat in my guest room," Fran wrote, "I poured out my heart to God and suddenly felt Jesus' presence. He was 'sitting' in the chair across from me and said to me, 'I know. I know.'"

Greater Memphis Habitat eventually found acreage in Rossville for a subdivision of fourteen houses. They also applied for and received a fifty-thousand-dollar grant from the National Presbyterian Church. That money, a fabulous sum in the early eighties, was used to pave the road and put in septic tanks. Habitat had a positive impact on that community and on Fran.

An "Unbelieving Socialist"

Malcolm Story, a self-described "unbelieving socialist" from England, did not come to God through his work with Habitat for Humanity, but he did make a keen observation concerning his involvement in the work. Malcolm started volunteering with the Lafayette, Indiana, Habitat affiliate in 1992. Over the years, he has worked faithfully with a group of retirees that became known as "the Magnificent Seven." He actively participated in the affiliate's building growing from twenty houses to more than fifty homes.

Malcolm says the following about his longstanding commitment to Habitat: "What is expressed by many—in my opinion wrongly—as a Christian spirit in participating in work to create a better life for those less fortunate than themselves, to me is answering a social conscience. But whatever final result is achieved, I get a sense of satisfaction in giving my time and efforts to Habitat for Humanity and in the company of such wonderful companions." And then he added this interesting note: "It could be said, when asking why an unbelieving socialist from England is drawn to work for Habitat for Humanity every year that it is 'the Lord moving in mysterious ways.'"

The Lord surely does move in mysterious ways to perform His wonders in Habitat for Humanity. The ministry has been from the beginning, and continues to be, a powerful vehicle for bringing people together and bringing many of them into a personal encounter and relationship with God.

All I can say is thanks be to God for this "more-than-houses" aspect of the work.

CHAPTER 11

Building Houses and Hope with Prisoners

This program is good because everybody wins.
I win because I am gaining skills I can take out into the world with me.
The people who receive the homes we build win because they get a
house that they might not have been able to afford otherwise.

Our instructor wins because he is passing on his knowledge to young
men making them better people and also giving them skills they can use
for career opportunities.

—CHRISTOPHER, INMATE HABITAT PARTNER

Big tears had welled up in Franco's eyes and started to trickle down his cheeks. He was a teenage inmate at the St. Charles School for Boys in St. Charles, Illinois. I was visiting that youth facility because the boys were building walls for Habitat houses. I want to tell you about Franco and about the exciting program at the youth center in St. Charles and in many other locations throughout Illinois and in several other states and Canada.

In 1904, the St. Charles School for Boys became the first state juvenile detention facility in the nation. In 1995, this facility, now called the Illinois Youth Center at St. Charles, very fittingly became the first juvenile detention facility to begin a partnership with Habitat for Humanity. Actually, the partnership was and still is a three-way relationship involving the Illinois Department of Corrections, Lutheran Social Services of Illinois (LSSI), and Habitat for Humanity. The first project of this partnership involved sixty teenagers putting together walls for a Habitat house, which were then transported to Americus, Georgia. I was privileged to participate in dedicating that house when it was finished in late 1995, and so was the director of LSSI. The partnership worked like a charm. A new branch of Habitat was a success.

The following spring, I was in northern Illinois on a speaking tour when I had the privilege of meeting the young men who built the walls for the Americus house and who had since then built many more walls for several Habitat houses, mostly for affiliates in Illinois. That was where I met Franco.

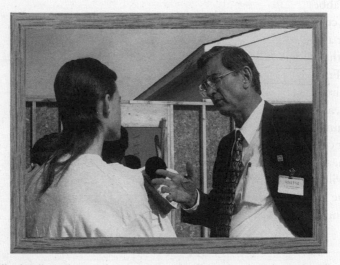

Millard Fuller talking with an inmate at the Illinois Youth Center in St. Charles.

Dick and Florence Nogaj, founders of the DuPage County, Illinois, Habitat for Humanity affiliate were also with me. At the prison, we were taken to an area outside a workshop. The boys were standing in a row, each one wearing a Habitat T-shirt. As I shook hands with almost all of them, they were smiling broadly. It was obvious that they were pleased to be a part of the Habitat program and very happy to be getting so much attention. We toured the workshop, and it was apparent that these young workers were very proud to show me what they were doing. I was impressed. During the program that followed, I expressed my gratitude for their work, offered some words of encouragement, and thanked the Department of Corrections and LSSI for making this unique partnership possible.

As I mingled with the crowd afterward, sipping punch and eating some crackers, I felt someone tap me on the shoulder. I turned around and faced a boy of maybe sixteen years of age. He was clearly nervous.

"Mr. Fuller, my name is Franco. I want to tell you how much we appreciate you taking time to come to visit us." He paused. "I've messed up my life. But one day, I will be released. When I get out I'll know how to build a house because of the skills I've learned in working on this project with Habitat for Humanity."

By this time, Franco's big tears had trickled down his face. "Thank you again, Mr. Fuller. I'm going to make something out of myself when I get out, and you've helped make it possible for me to do that."

I grabbed Franco's shoulders and pressed hard. As I looked him in the eyes, I could feel tears welling up in my own eyes. We stood there speechless for a long moment. Then he turned and walked away.

As the Nogajs and I left the prison that day, I couldn't quit thinking about Franco. I have no idea why he was there. I don't know what crime he committed. But I do know this—Jesus said that we should visit those in prison, and that a visit to "the least" person in prison is just like a visit with Jesus himself.

Franco certainly was in the category of "the least," but God's love extends to him, and it filled me with great joy to realize that I had had the wonderful privilege of being of help and a blessing to that young man. I was thrilled that the Habitat ministry had penetrated the walls of that prison and was touching lives inside in such a profound way.

New Frontier

This new frontier in Habitat for Humanity's work was inaugurated as a result of a meeting I had with a special man in Chicago in November 1991. Linda and I were in the city to attend the annual meeting of LSSI. I was the featured speaker that evening, and I was also privileged to be the recipient of their Amicus Certus award. After the program ended, I was visiting with people in the audience. Jack Nordgaard, Illinois state director of the Lutherans prisoner and family ministry, approached me and asked, "Why don't you have a program for prisoners to work in Habitat?"

"Because you haven't started it!" I replied.

Jack accepted that challenge. In the coming months, he got busy in organizing a program for prisoners to work with Habitat affiliates.

The first venture Jack launched was with the Kankakee Minimum Security Unit for Women and Pembroke Habitat for Humanity in St. Anne, Illinois. In late 1993, thirty women from the unit were taken to a construction site in St. Anne. Their task was to renovate a house for a needy family. The women worked faithfully and diligently for eight months. The work was finally finished in May 1994, and the house dedicated and turned over to a grateful family. In October of that year, the Kankakee unit formalized a Habitat group in the facility.

By late November 1996, three Illinois prisons were building Habitat walls in addition to the ongoing work by the Kankakee unit. Those prisons were Western Illinois Correctional at Mount Sterling and the Illinois youth centers at St. Charles and Harrisburg.

The Department of Corrections agreed to provide the skilled personnel to work with the inmates to make the walls for Habitat houses. Some of the construction training would be done on a contract basis with area colleges. Those walls would then be taken to Habitat building sites. LSSI would provide the funding for purchasing the building materials. Habitat for Humanity affiliates would prepare the foundations and coordinate all aspects of the construction on site and be responsible for dealing with the homeowner families in regard to selection, sweat-equity requirements, counseling, collecting monthly payments, etc.

In May 1996, Jack Nordgaard retired as director of the prison and family ministry of LSSI, but he remained vitally involved in the growing work. Jack was replaced by a dynamic woman, Jane Otte. She has continued the strong support that Jack inaugurated and has also helped export the idea of prison partnerships to other prisons in Illinois and to other states.

As of early 1999, the Prison-LSSI-Habitat partnerships in Illinois had accomplished the following:

- Sixty Habitat houses built by prisoner partners

- Thirty-seven of those houses built by incarcerated juveniles

- Twenty-one local Habitat affiliates participating in the program

- Nine affiliates have done more than one house with a prison partner

- Six correctional facilities building prefab units for Habitat affiliates

- well over five hundred prisoners now have construction skills necessary to build walls and other component parts for Habitat houses

Prison Fellowship's Early Work

Actually, prisoners had already been working on some Habitat houses. In Mississippi, for example, prisoners had worked at various construction sites.

We had also worked with Chuck Colson's Prison Fellowship ministry in several locations. In 1986, Chuck, along with a dozen prisoners, worked at the Jimmy Carter blitz project in Chicago. But no formal program for work with prisoners was ever inaugurated until the work was launched in Illinois.

In early 1998, the adult prison at Taylorville, Illinois, Taylorville Correctional, joined the Habitat-LSSI partnership. In the first six months, the men at Taylorville completed walls and trusses for six houses. One set of walls and trusses went to the Habitat affiliate in Quincy, three were sent to east St. Louis for a June blitz-build, and one set was trucked to Houston, Texas, for the Jimmy Carter one-hundred-home blitz that was also held in June. Twenty-five prisoners from the Prison Fellowship's InnerChange Freedom Initiative built the house in Houston utilizing the prefab walls and trusses from Taylorville. The same inmates, incidentally, helped with the prebuild work for the blitz-build in Houston.

More Than Walls

While most work in Illinois prisons has involved constructing walls and trusses, inmates from the Western Illinois Correctional Center built kitchen cabinets and utility sheds for Habitat affiliates. At the Illinois River Correctional Center in April 1997, exterior and interior walls were built to specifications for a disabled child of a McDonough County Habitat-selected family.

Jacksonville Correctional Center inmates, as a project of their construction class, helped erect Habitat houses on site in the local community in 1995 and 1996. And in September 1996, students from the Hanna City work camp construction operations class set up and poured footings, established drainage, and poured concrete walls for two houses sponsored by the Peoria Habitat affiliate. The participating affiliates, correctional officials, and the inmates were very pleased, indeed, excited about the prison partnership in Illinois.

No Longer Depressing

The Reverend Buck Jones, president and founder of the Habitat affiliate in east St. Louis, visited the Taylorville Correctional Facility several times while the walls and trusses were being made for his blitz. "Usually, when I go to prisons, it is depressing," he said. "But at Taylorville, they're really teaching these young men skills. And these inmates tell me, 'Reverend, it feels great

An inmate at Illinois' Taylorville Correctional Center meticulously measures lumber for a Habitat house.

to be able to put something back into the community and to do something good for someone.' Several have said it gives them a sense of worth. And the quality of the men's work is excellent."

Jim Ecklund, executive director of Habitat for Humanity of Knox County, headquartered in Galesburg, wrote the following about the partnership. "I am glad to let other Habitat affiliates know about our experience working with LSSI and the Illinois Department of Corrections—to have prisoners build the walls for our most recent home. We have nothing but good to say about the partnership. The center not only built the exterior and interior walls, but sheathed the outside walls and installed the windows, so that when delivered on site, it took our crew only a couple of hours to put them all in place, ready for trusses to be put on. We want to work with LSSI to do this again, and we encourage other affiliates to consider this partnership as another way we can bring people of all backgrounds together in our shared ministry of building simple but decent homes for God's people in need."

Building Hope

Robert Montgomery, president of Helping Hands of Benton Habitat for Humanity, was totally pleased with the prison partnership: "We received well-built prefab walls and trusses. The walls were prewired with wall box receptacles and switches, all in the proper place. Sheeting was on the outside walls. The interior walls were also prefabbed, which greatly reduced the time to set the walls and trusses. We met with many of the young men from the youth center and were greatly surprised how much they had learned and how much pride they felt in doing something for someone else."

Charlotte Jensen, executive director of Habitat for Humanity McDonough County, headquartered in Macomb, had this to say about the prison program they shared with the Illinois River Correctional Center at Canton. "Prior groundwork by LSSI and the success of the prison ministry in other correctional facilities made our coming together painless. Walls and houses are the visible signs of this ministry but something even greater is being built—hope!"

Homeowners Meet Prison Partners

Homeowners who receive the Habitat houses that are partially built in prisons often go to the correctional facility to meet the inmates who built their walls and trusses (or other parts of the houses) and express their appreciation to them. Clarence Golden, a recipient homeowner in Carbondale, went to the Illinois youth center at Harrisburg to meet and thank the young men who built the walls for his house. Clarence shared that he also had made mistakes as a young man. He told them how he had long dreamed of owning his own home. He encouraged the young men to continue learning construction skills so that they too could build their own home one day.

One homeowner cried when he met the inmates who had built his home. He explained his emotion by saying that he was not that much older than the juveniles himself. He had had a similar background and said that he could have been in prison himself.

Holly Reynolds of Round Lake Beach went to the youth center in St. Charles to thank the young men for their part in making her house possible. She told them about her situation before getting the house. "It was eviction notice after eviction notice," she said. "Eventually, I said a few prayers and found out about Habitat for Humanity and at that point I was determined that, yes, I'm going to build a home and provide a family atmosphere for my kids." She said hers was a story of hope and faith because she stuck with her dream.

Then the young men at the center did something very surprising—they gave Holly a plaque that read "Home Sweet Home." They also presented her a gift of seventy-five dollars to spend any way she chose. That is pride in action demonstrated by young men with new hope.

First-Year Accomplishments

Lawrence Jaeckel, assistant superintendent of programs at the Illinois youth center in St. Charles, wrote Jack Nordgaard in November 1995 about the emerging partnership. He reflected on the accomplishments over the first year of the cooperative venture.

First, he listed the needs that were being met for the young inmates:

1. To learn marketable skills such as building construction trades.
2. To be involved in success experiences that would help improve their self-esteem and self-worth.
3. To learn to incorporate into their values the personal reward inherent in helping others.
4. To have as many positive links to the community as possible to provide guidance and support once they were released from our facility.
5. To make amends to the community through a positive contribution.

He went on to point out that the partnership operates without additional tax dollars, helps provide affordable housing for families in need, teaches young men marketable skills and the values inherent in helping others, and provides a viable follow-up resource for support services to youth in the community once they are paroled.

Jaeckel concluded his letter by thanking LSSI for being the catalyst to bring the partnership into existence and for providing the glue that helps hold all the pieces together. "It is an excellent service for the youth assigned to the facility and to the public," he said. "Again, thank you for helping us provide the best available experiences to our youth through this program. I view it as a truly positive investment in their future."

Productive and Meaningful Contributions

Early in the following year, Odie Washington, director of the Illinois youth center at Harrisburg, wrote to Jack Nordgaard to express his appreciation for the partnership:

This project does more than just provide houses to people who desperately need them. It provides our students excellent hands-on building experience that we might otherwise not be able to offer. It also provides them with the opportunity to help other people who are in need.

This is probably the most important experience we could provide them. The students in the program begin to understand they can make productive and meaningful contributions to other people in our society. They express this when they talk about what they are doing, and they show it in the quality of their work.

A few months later, in July 1996, William D. O'Sullivan, warden of the Western Illinois Correctional Center in Mount Sterling, wrote to Jane Otte about the program:

It is remarkable just how involved the students become in the home projects and the sense of satisfaction they feel each and every time a project is completed. The Habitat for Humanity projects enable students to make a productive contribution to society which creates a win-win combination. On behalf of Western Illinois Correctional Center and MacMurray College, I want to thank you for helping to create a winning relationship between LSSI and our institution.

Inmates' Praise

The inmates are especially effusive in their praise of the prison partnership with Habitat for Humanity. Chris, a teenage inmate at the Center in St. Charles, wrote the following:

When I started building houses for Habitat I was seventeen. Now I'm eighteen. I feel that I have accomplished a lot. This program has given me self-confidence because it has shown me that I don't have to be a loser any more. The program has taught me that I can help other people and that I'm good at something. Working with Habitat has taught me many different skills, like reading blueprints, building walls and cabinets, and how important teamwork is. I know these skills will be very useful in the future for all of us involved in the program. I also know that this program has helped at least three people get jobs.

Matthew, another inmate, wrote, "I was convicted of burglary and sent here in August 1997. I was just going to do my time and nothing else, but someone told me to do otherwise. That person was the Lord. I prayed, worked for my GED, and became involved in an auto mechanics class. Unfortunately, the auto mechanics class didn't work out so I enrolled in the wood shop class. Just last week we finished a set of kitchen cabinets for a disabled individual to be installed in a Habitat house. I had never done anything like that before and it made me feel pretty good."

Everybody Wins

Another inmate, Christopher, wrote about how the program was a positive experience for everyone involved:

When I first started, I had barely any knowledge in woodworking or carpentry. Since I have been in this class I have learned a lot of valuable skills—skills such as making accurate measurements and how to safely use dangerous tools. And through this gain greater self-discipline and focus. The wood shop class has also given me a chance to be involved with Habitat for Humanity.

This program is good because everybody wins. I win because I am gaining skills I can take out into the world with me. The people who receive the homes we build win because they get a house that they might not have been able to afford otherwise. Mr. Hart, our instructor, wins because he is passing on his knowledge to young men, making them better people and also giving them skills they can use for possible career opportunities.

New Frontier Broadens

As the prison partnerships continued to grow in Illinois, other prison partnerships were emerging in other places. In 1994, in Broward County, Florida, the local Habitat affiliate, Habitat for Humanity of Broward, headquartered in Pompano Beach, started working with inmates.

The Broward sheriff's office (BSO) had just initiated a boot camp for young male offenders. Lew Frazar, executive director of Broward Habitat, met with the lieutenant in charge of the program to see about using the inmates to build Habitat houses. Lew said that particular effort didn't work out, but the contact got some creative juices flowing. Within a month, he reported, the affiliate was getting crews from BSO a couple of days a week to work on various yard and lot maintenance projects. Soon thereafter BSO decided that Habitat was to be a permanent assignment for a crew of five to seven inmates. The first crew included an electrician and a tile setter.

In the years following, the program evolved from a very informal start to a formal program. Inmate crews come five days a week to help build Habitat houses. They are supervised by a deputy who knows so much about the work that he sometimes acts as a supervisor-leader. Eventually, a second deputy was assigned to the crew.

As the program enters its fifth year in 1999, it is stronger than ever. Kit Raines, director of development for Habitat for Humanity of Broward, serves as project manager. She is most enthusiastic about her work with the inmates and is particularly pleased with the impact the program is having on the men (and sometimes the women who now participate).

Instruction Beyond Construction

In 1998, the program was expanded to include not only training and work experience in home construction, but also instruction on how to complete applications for employment and how to interview for jobs. And the inmates receive substance-abuse training twice a month on the job site in addition to that received in the jail. The cost for this training is covered by a grant from the Broward County Commission on Substance Abuse.

Some inmates are referred upon release to the AFL-CIO Building Trades Council Apprentice program. As of early 1999, eight former inmates had been accepted. Other former inmates have gotten jobs directly in the construction industry upon release.

Both Lew Frazar and Kit Rains say that the most important thing about the program is the intangible benefits to the inmates. They frequently get comments such as:

"This is the first time in my life I ever did something for someone else."

"I'm going to volunteer when I get out."

"I appreciate the opportunity to learn."

"I really enjoy doing something for someone who is poor like me."

"At Habitat I am treated like a real person, not just a prisoner with a number."

Lew says that the inmates—by working alongside regular volunteers, college students on spring break, church groups, and corporate volunteers—get to see a side of life they have never seen before—God's people putting their faith into action. In May 1998, Lew left Broward Habitat to become executive director of Tampa Habitat for Humanity. There, he initiated conversations with Jan Bates, inmate programs manager of the detention department of the Hillsborough County Sheriff's office about involving inmates in helping build Habitat houses. Plans are in place to utilize inmates at the Falkenburg Road Jail to build walls and trusses for Habitat houses.

Prisoners lifting a truss for a Habitat house built in Tallahassee, Florida, sponsored by the Florida Homebuilders Association. This house was built in less than twenty-four hours on April 7, 1999.

Building Instead of Breaking into Homes

In the Florida panhandle, Pensacola Habitat for Humanity has been working with Pensacola Boys Base since July 1997. The Boys Base is a juvenile detention facility. The teenagers who are incarcerated there have helped build several Habitat houses during the first two years of the partnership. The young men learn vocational skills and the importance of teamwork, and they build self-esteem. A unique and very special benefit of the program is that the boys, many of whom had been breaking into homes, are now giving back to the community by building homes.

Up the coast in North Carolina, inmates in the Forsyth Correctional Center in Winston-Salem have been building walls for Habitat houses since 1996. That partnership between the center and Forsyth County Habitat for Humanity was launched by Roger Jones, a board member of the Forsyth Jail and Prison Ministry and an active volunteer with Forsyth Habitat. A total of five sets of walls had been built at the correctional center by the end of 1998 and plans are in place for building at least another set of walls in 1999. The inmates have even gotten involved in some of the fund-raising. One inmate, who earns seventy cents a day for his work in the prison, donated eight dollars.

In Sioux Falls, South Dakota, prisoners are helping with on-site construction of Habitat houses. I was there in 1997 and visited with some prisoners who were putting up dry wall in a Habitat house.

A Mailbox and Address Numbers

In Canada, work with prisoners is emerging in Brandon, Manitoba. The Brandon Correctional Institution has been providing inmates to the local Habitat affiliate for the past few years. Ray Heppele, a volunteer with Brandon Habitat, is employed as a jail guard at the Brandon Institution. He initiated the program.

In 1996, Wilmer Martin, president of Habitat for Humanity of Canada, and the Right Honorable Edward Schreyer, former governor-general of Canada, and his wife, Lily, were at a blitz-build in Brandon. Each morning Ray would bring seven men from the institution to work on the house—the first one built by the affiliate. At the end of the week, the men took a collection among themselves and bought house numbers and a mailbox for the family. These gifts were presented at the dedication service. The men embraced the single mother as tears flowed freely both from the overjoyed homeowner and the inmates.

Since that first house, the Brandon affiliate has built three more. Each one has been built in part by volunteer teams from the Brandon Correctional Institution. The men are good workers. One in particular, though, stands out. He is a native Canadian who has helped on a number of houses.

Work on one house was running a bit behind schedule. After the men had been taken back to the institution for the night, this man convinced Ray to take him back to the house so he could do additional work to ensure that the house was finished on time.

First Official Prison Habitat Chapter

In Abilene, Texas, prison volunteers from the Robertson Unit blitz-built a Habitat house in 1994. Meanwhile, in Dallas, Jim Pate, executive director of Dallas Habitat for Humanity, and Carol Freeland, member of the

Habitat International board, were working out a partnership with the Hutchins Correctional Unit to become the first official prison chapter of Habitat for Humanity. Inmates from Hutchins started working on Habitat houses in 1995.

Carol wrote about the beginning of the relationship and about a most interesting encounter in December 1995 with an inmate at the site of the first Habitat house they worked on. "Building relationships, breaking down barriers, and bringing people together happened with the first prison chapter organized by Habitat for Humanity. Inmates of the Hutchins State Jail in Dallas were skeptical at first that Habitat and the Texas Department of Criminal Justice could even use a group of prisoners to do anything positive—certainly not something as important as building a house! But they could and they did. Twenty inmates helped complete and dedicate a house in south Dallas in time for a single father of five children to be in the new home for Christmas.

"In putting up the walls for the house, other 'walls' fell as the prisoners got to know each other better, and as they got to know their guards and Habitat staff, volunteers, and the homeowner," she said. "The inmates worked with their guards in learning how to put on roofing. They discovered previously unknown talents in those guards too.

"At the dedication of this first of many Habitat houses built by the prisoners, I stood next to one of the younger men who had worked on it. He was both excited and nervous. His low self-esteem was 'printed' on his forehead for all to read. He and I had spent a fair amount of time together. I asked him what he was feeling, and he responded, 'Happy.'

"'What did you work on?' I inquired.

"'Oh, not very much that was important like the roof. I just mainly did that step over there.'

"'You mean this step?' I said as we walked together to the front porch.

"'Yep,' was his head-down reply.

"'So, if I understand you correctly, you built this cement step, which is the first step into this house. Everyone who enters will put their foot on your step. Is that correct?'

"His head came up and he looked me in the eye. A big smile creased his young face. 'That's right. Yep. That sure is right!'"

On July 24, 1996, a formal covenant was signed with the Hutchins State Jail Habitat group making them the first official prison chapter of Habitat for Humanity International. (See Appendix III for a copy of that covenant.) In the coming months, more inmates joined the program, and they continued to help build Habitat houses. As of early 1999, more than eighty inmates had participated in the growing partnership with Dallas Habitat for Humanity. These inmates had worked on a total of thirty-two houses and had total responsibility for three blitz-built houses.

Jim Pate says that the inmates now work with Dallas Habitat three or four days a week. He said all of the men are volunteers, even though they are screened by the administration. There is a waiting list to join the Habitat crew. The inmates are all serving a sentence of two years or less and must have a four- to six-month clean record in the unit (no fights or disciplinary actions) before they can be considered for the Habitat crew.

Reconciliation Outreach Partnership

In addition to building Habitat houses, the inmates have also worked with Dallas Habitat for Humanity on several other exciting projects, such as a two-story house built in partnership with Reconciliation Outreach, a non-profit organization with a ministry to ex-offenders. This facility now provides housing for eight men. The inmates also worked on a complete renovation of an old building to house the Dallas Weed and Seed program. This initiative combines U.S. Department of Justice staff, DEA staff, and local police, fire, and code enforcement personnel to "weed" neighborhoods of bad influences and "seed" with new or rehabilitated housing and businesses. Dallas Habitat for Humanity received an award from the Greater Dallas Crime Commission for their work on the Weed and Seed project and for their work in building decent, safe, and affordable housing. The Hutchins chapter also worked extensively on the racking, displays, and finishing work on the warehouse facility that houses the Dallas Re-Store.

Because of their outstanding work, the men were recognized at an awards luncheon on April 10, 1997, and presented the 1997 Outstanding Volunteers award by the Volunteer Center of Dallas County. The next day the *Dallas Morning News* ran an editorial about the inmates who received the prestigious award:

> Dressed in white prison uniforms, the men provided a jarring contrast to the business people who attended the annual volunteer center luncheon. The inmates' broad smiles dispelled any apprehension the prison garb may have created.
>
> The prisoners relished their recognition as the county's top group of volunteers. And well they should. The prisoners worked many hours

with Dallas Area Habitat for Humanity, building homes in southern Dallas. Along the way, they discovered the blessings of serving others.

Several inmates say they plan to use their home-building talents once they get out of jail. Perhaps more important, they regained a sense of self-worth, which probably was missing since their lives took a wrong turn.

The Habitat for Humanity Chapter at Hutchins State Jail is the first of its kind in a detention facility. But it won't be the last. The program's success already has caught the attention of other wardens looking for ways to help rehabilitate their inmates.

Texas Criminal Justice Board member Carole Young from Dallas and local Habitat for Humanity executive director Jim Pate deserve credit for making this idea succeed. Ms. Young, who conceived the plan with Habitat International board member Carol Freeland, is convinced that such programs can show inmates how to restore their lives.

Along with Mr. Pate and Hutchins State Jail warden Elvis Hightower, Ms. Young should feel as proud as the beaming inmates. They all took a risk. It paid off.

In September 1997, I was in Dallas for Building on Faith week. An excited group of men from the Hutchins Correctional Unit was there as an integral part of the work force. I talked to them and expressed my gratitude for their partnership with us in the ministry of Habitat for Humanity.

Texas Board of Criminal Justice Proud

In October 1998, I received a letter from Allan B. Polunsky, chairman of the Texas Board of Criminal Justice, praising the growing partnership with Habitat for Humanity:

On behalf of the Texas Board of Criminal Justice and Executive Director Wayne Scott, I want to extend our continued commitment to Habitat for Humanity. As you're aware, the partnership established with the Texas Department of Criminal Justice (TDCJ) has been recognized as a corrections "best practice" by the American Correctional Association (ACA) this year. We are proud to be involved with this excellent community service program recognized for helping Texans realize the dreams of homeownership. We are also proud to be able to provide our offenders an opportunity to give back to their communities.

In 1997, the agency expanded the program. First at a maximum-security prison, offenders who were confined to a facility constructed a house in modular sections. These activities took place within the facility's compound. Next, offenders built cabinets for future installation at homes and crafted wooden toys and doll houses to be sold to raise funds for Habitat. TDCJ also helped local affiliates recycle donated building materials by demolishing old structures; the offender volunteers even disassembled a movie set.

The demonstration project at the San Antonio ACA winter conference in January received positive media coverage and resulted in interest from several states in developing similar programs.

The success of our partnership has surpassed our expectations; the agency is committed to continuing programs that increase the self-esteem of offenders which is key in the chain of our rehabilitation efforts.

Trucker's House

Jim Pate reports that a number of the former inmates have visited the Habitat offices after their release to express gratitude. A few, he says, are now long-haul truckers, and they stop to visit when passing through Dallas. Jim has recruited one of those truckers to help organize a Trucker's House to be built in partnership with trucking companies and trucker ministries during the 1999 Building on Faith week.

In all, a total of twenty-one prisons in Texas are working with fourteen Habitat affiliates, and more prison partnerships are in process of formation in other locations in Texas.

Working with Wisconsin

In Wisconsin, Jan Nigh, affiliate support manager for Habitat affiliates in the state, worked with state representative Clifford Otte and Governor Tommy Thompson to create a partnership between the Wisconsin Department of Corrections and Habitat for Humanity. This partnership was announced early in 1998. Governor Thompson said on that occasion:

Through the combined efforts of the Department of Corrections, various technical colleges, and Habitat for Humanity, we have the opportunity to provide low-cost housing to deserving families across the state. This partnership not only enables inmates to give something back to the community, but also allows them to learn valuable skills and gain a more positive attitude toward people and society. I commend Habitat for Humanity for its contribution to the citizens of Wisconsin and across the country.

The first project of this initiative was at the Columbia Correctional Institution in Portage. On April 18, 1998, ten inmates began building the walls for a Habitat house to be erected by Greene County Habitat for Humanity in Monroe. The inmates who worked on the project were selected from the building services program. They worked diligently to complete the walls in three weeks. In the process, they learned work skills and the ability to construct the framing of a house from beginning to end. The inmates also had a lot of pride in the project, knowing that the home would be occupied by a family in need and that it gave them the opportunity to give something of significance back to the community. The walls were delivered to the Habitat building site in Monroe in early May. On the ninth of that month, Governor Thompson, Representative Otte, Jan Nigh, and other volunteers erected them. In the following weeks, other volunteers finished the house. It was dedicated in November, and the family moved in to enjoy Thanksgiving and Christmas in a wonderful new home.

Soon after completing the walls for the Monroe house, the inmates at Columbia Correctional started a set of walls for a Habitat house in Milton to be built by Rock County Habitat for Humanity. That house was built over the summer and fall and then also dedicated in November. Funds for materials for the walls built by the inmates at Columbia Correctional came from a ten-thousand-dollar grant from the Patrick and Anna M. Cuday Fund of Milwaukee. Judith L. Borchers, executive director of the fund, said in a letter to Jan Nigh: "Congratulations on your fine work, for which we are happy to provide this support."

Cabinets, Flowers, Shrubbery, and Much More

As the walls were being built by Columbia Correctional, inmates in other correctional institutions were also busy with Habitat work. At the Fox Lake Correctional Institution in Fox Lake, Wisconsin, for instance, kitchen cabinets were built for Habitat houses in Hancock, Milton, and Monroe. In Green Bay, at the Green Bay Correctional Institution, inmates assembled cabinets for Greater Fox Cities Habitat for Humanity in Appleton. The Oshkosh Correctional Institution planted and cultivated flowers and shrubbery for Habitat houses in Appleton. That institution also plans to build eight sets of walls and eight sets of cabinets for Habitat houses in 1999, in addition to continuing to grow flowers and shrubs for Habitat houses.

As I write this, a brochure is being planned to market the services of fifteen state correctional institutions to Habitat affiliates in Wisconsin. Those services include making storage sheds, mailboxes, house address numbers, concrete steps, ramps for disabled people, wooden toys for fund-raising, crochet and knitting for clothing and bedding for fund-raising, recognition plaques, ceramic items (such as lamps, vases, and Christmas decorations) for fund-raising, and architectural designing in addition to the services that are already being provided.

Changing Inmates' Lives

A steering committee has been organized to coordinate this growing partnership between the correctional institutions of Wisconsin and the various Habitat affiliates, currently numbering thirty-five. Jeff Endicott, warden at Columbia is chairman of the committee. Chuck Barnum, a dedicated Habitat volunteer, is co-chairman to handle the affiliate side of the partnership. All Habitat affiliates in the state and all correctional institutions hopefully will be involved in the near future in this pioneering venture of building houses and a whole lot "more."

The program is making a difference in so many ways and is a tremendous blessing to the affiliates. But perhaps the greatest difference the program is making is in the lives of the inmates. Consider what two inmates have said about the experience of building for Habitat for Humanity houses.

James Jardine of the Columbia Institution said, "I have been locked up for five years. I enjoy working with my hands. I took a building construction course and graduated with highest honors. When the chance came to build a Habitat house I was the first to sign up. This is the only chance I'll have to put my training to work. I want to thank the Habitat people from Greene County for believing in us and for giving us a chance to show everyone that we're not as bad as some people want to believe. I look forward to building more houses."

Teddy Chambers, also at Columbia, was equally effusive in his praise of the program. "In a place where I feel like I'm being systematically trained to hate and to hate life, I would like to thank Habitat for Humanity for giving me the opportunity to show that I am a caring person. Thanks for the experience. I've learned a lot. My only regret is that I will be sent away and will be unable to participate in the next Habitat project." Teddy was transferred to a prison in Texas, and to his delight, there was a Habitat partnership in that institution so he was able to continue his Habitat building.

And a heart-wrenching example of what Habitat's prison program can do for *any* inmate's heart, LSSI's Jack Nordgaard recently told me about John Whitehead. John is a prisoner who had just sent him a check for five hundred dollars, a donation to the prison partners program in the Midwest Region. His letter giving me John's address so I could thank him included this line: "The note should be sent to Mr. John Whitehead, Menard CU. The CU stands for Condemned Unit."

Prisoners at Saginaw Correctional Facility in Michigan stand with Michigan state Habitat director Ken Bensen (far right).

Michigan Working Hard

As the prison work was developing in Wisconsin, Ken Bensen, state director of Habitat for Humanity in Michigan was hard at work launching a program in that state. The first partnership was with the Saginaw correctional facility. Starting in May 1998, inmate volunteers prefabbed interior and exterior walls for five Habitat houses for four different affiliates—Saginaw Habitat for Humanity, Mid Thumb Habitat in Palms, Macomb County Habitat in Mount Clements, and Bay County Habitat in Bay City. Each affiliate got one set of walls except for Mid Thumb Habitat, which received two sets. The first walls to go up were in Saginaw. That happened on May 30, 1998, while a five-house blitz was underway.

John Perna, executive director of Saginaw Habitat, was most enthusiastic in his praise of the work done by the inmates. "Great program!" he exclaimed when I asked him about their work. "The walls were color-coded. Everything went together perfectly. The men really did a nice job. Very professional."

The Saginaw house was finished in early July. The other four houses were completed later in the year. Governor John Engler attended and spoke at the dedication of the house built by Macomb County Habitat for Humanity. Ellen Schippert, an architect in Palms, designed the houses. In fact, she came up with five sets of plans so that local affiliates could choose one or more designs. Ken Bensen worked out an excellent deal with Wolohan Lumber Company of Saginaw, whereby lumber was delivered to the institution and billed to the appropriate affiliate at a favorable price. Bruce Hitsman, project supervisor at Saginaw Correctional, works closely with Butch Urban at Wolohan to ensure a smooth-running operation that fulfills the needs of the Habitat affiliate. Saginaw Habitat is planning to blitz-build six houses in June 1999 as a part of the Jack Kemp one hundred and eighty house blitz-build. Saginaw Correctional plans to build the walls for ninety of those houses.

Ken Bensen is talking to the Michigan department of corrections to involve all thirty state correctional institutions in building walls and, in some instances, complete houses to be shipped out to the eighty-three affiliates in Michigan.

An Award Winning Program in Tennessee

In February 1999, when I was in Memphis, Tennessee, on a speaking tour, I learned firsthand about an exciting program in that city involving prisoners building Habitat houses. It is called Building for the Future.

Begun in 1993 as a small job training program for inmates of the county prison system, it has expanded into a much larger initiative involving on-the-job training in masonry, wall panel and roof truss construction, painting and construction finishing skills, including drywall. The program benefits Habitat for Humanity and other housing organizations.

Susanne Jackson, who has been with BFF from the inception, says that nearly nine hundred men and women were enrolled over the first five years and have been involved in building forty-four Habitat houses.

The corrections division, which has kept track of the released inmates who were a part of the program, has found strong evidence that these men and women tend to stay in construction and do not revert to a life of crime. Eddie Walsh, director of Building for the Future, said that one ex-offender even started his own company! Another former inmate was selected to receive a Habitat house for himself and his family.

On September 9, 1998, the Building for the Future program received a finalist certificate in the 1998 Innovations in American Government competition sponsored by the Ford Foundation and Harvard University's John F. Kennedy School of Government in partnership with the Council for Excellence in Government. The certificate cited the "novel, exciting, and highly successful approach" employed to meet pressing public needs.

Idea Whose Time Has Come

Michael Willard, director of Habitat International's program enhancement department, has responsibility for the prison partnership program. He has received inquiries from affiliates and prison representatives from Louisiana, Washington, Ohio, and many other states. So it is obvious that this whole concept of having prisoners work with Habitat for Humanity to build Habitat houses and to help the ministry in a number of ways is an idea whose time has come.

In the United States, more than 1.7 million people are incarcerated in local jails and in state and federal facilities. And the numbers are rising every year. It behooves society to help these inmates have good experiences in prison—learning and growing experiences—so they will have a better chance to succeed when they are released.

As this chapter reveals, the prisoners who have been involved with Habitat for Humanity have had good experiences. They have helped Habitat and they have been helped. The whole program has indeed been building houses and, wonderfully, a whole lot more for a segment of our population usually ignored, marginalized, and left too often hopeless.

As Charlotte Jensen, executive director, Habitat for Humanity McDonough County, prison partners with Illinois River Correctional Center, says, "Walls and houses are the visible signs of this ministry, but something even greater is being built—Hope."

CHAPTER 12

Bridging Religious, Political, and Cultural Differences

At our Habitat functions, we include all singing groups from various churches in the area. The poor and the rich now mix freely without feeling bad about the other group. Habitat for Humanity has managed to bring these different groups together.
—WAVESON HAMANI HAMUCHANKWI, HABITAT LEADER, ZAMBIA

A common meeting ground. That, from the inception of Habitat for Humanity, is what I envisioned this ministry to be for all faiths and persuasions. This hammer-and-nail ministry brings together all sorts of people from all walks of life.

Foremost among my desires is the ministry's potential of being a bridge for Christians—all members of the Christian faith in all its various denominations. Christian churches are so divided, and the divisions are on so many levels. There is high church and low church, liberal theology and conservative theology, charismatic and noncharismatic, evangelical and mainline, Catholic and Protestant. There is no agreement on communion: how often to take it or what it means. Churches can't agree on how preachers should dress, much less what gender preachers are permitted to be, or even what to call "preachers"! The list of things that divide churches goes on and on. I have always thought it was shameful that the Christian family could agree on so little.

My fervent hope has been that this ministry would be something that a broad range of people of faith could agree on. I wanted to give my brother and sister Christians a ministry that all could feel good about and agree on. And what better idea than providing one of the most basic of human needs—shelter—and utilizing a tool that has so much significance to the disciples of Jesus—the hammer. After all, Jesus employed a hammer in Joseph's carpenter shop in Nazareth when He was a boy. It was a hammer that was used at the end of His earthly ministry to nail Him to the cross at the moment of deepest sacrifice for us, something every church accepts no matter what terminology it uses to describe the sacrifice. And it is universally accepted in all Christian circles that the clear mandate of the Christian gospel is to love. As Jesus was about to leave His disciples at the end of His earthly ministry, He instructed them to love one another. In addition, He made it clear that in giving the very basic things like food, water, clothing, and shelter, we are sharing love. Matthew 25 states that when we invite a stranger in, it is as if we have invited in Jesus himself.

So there's obviously broad agreement among most Christian denominations of the core concepts and theological basis of Habitat for Humanity. Most denominations also agree on the practical aspects of the work, namely self-help, that sweat equity offers not a giveaway, but rather a way to help people help themselves. Selling houses to low-income people at no profit and no interest makes it possible for them to afford a home. The families are given a wonderful opportunity to improve themselves, but their dignity is not assaulted nor taken away. All of this is attractive to a wide array of Christian denominations, so churches are able to embrace the Habitat ministry even though they might differ strongly on other subjects.

Other Bridges—"Cosmic Connection"

Beyond the Christian family, I desired that others too would be invited to join in the great task of eliminating poverty housing. After all, I reasoned, one does not have to be a Baptist, Methodist, Catholic, or adherent of any faith to help build a house for a needy family.

Many Jewish individuals and Jewish groups, including synagogues, have long embraced and supported Habitat for Humanity. My good friend Will Leventhal of Malibu, California, was one of the early strong supporters of this ministry. He says of his partnership with Habitat:

Involvement with Habitat for Humanity has always been so special for me because I have always been welcomed by and felt the sincerity of the folks I've met. Each time I've been blessed to take part in Habitat activities—whether building in Americus or south central Los Angeles or walking to meetings in Indianapolis or Kansas City, the spirit and diverse people involved have been wonderful and energizing.

I was able to work on the initial Jimmy and Rosalynn Carter work project in New York City in 1984 and later, in 1988, I tried to explain to the student Hillel group at Washington University in St. Louis how the "hands-on" group building experience was both personally uplifting and a great "mitzvah" (good deed).

There are many good deeds—food drives, clothing donations, tutoring, etc., done by Christians and Jews in the United States, but what makes Habitat so unique and fun is that people come together for the common good (what Dr. Martin Luther King Jr. termed the "Beloved Community"), you work up a good sweat, and at the end of each day you see the immediate fruits of your collective labors—and know that it is a seed of happiness and strength for the family who is working there with you!

That feels great—plus people working alongside you are always so "up," and the energy is magnetic and geometric.

Finally, for me a Jew—having been to Israel and lived on the Jordanian border at Kibbutz Ashdot Ya'acov (at the confluence of the Jordan and Yarmulk Rivers) building basic housing—as they did in the early Kibbutzim and as Jesus possibly did—provides an uplifting cosmic connection.

In Spite of It

People of other faith persuasions and individuals of no professed faith have likewise supported the work. I well remember a letter from a woman in Florida. She was upset about all of the religious language that was used in our literature. "My parents freed me from religious myths," she exclaimed. "I know the Bible is bunk and that Jesus was a hoax, but I think what you are doing in building houses for poor families is wonderful. So enclosed is my latest contribution of $100, but do understand that I send it not because of your 'Christian witness,' but in spite of it!"

I wrote back and thanked her for the letter and the gift. I invited her to visit us at her first opportunity.

The late Butterfly McQueen, movie actress of *Gone with the Wind* fame, was a Habitat contributor. I met her in New York City in 1985 at the Jimmy Carter blitz-build. She corresponded with me after that on several occasions. When she wrote, she always stated that she was an atheist, but a contribution came with every letter.

A Mosque and a Methodist Church

In September 1997, Dallas, Texas, Habitat for Humanity had the honor of building the milestone sixty thousandth house during Building on Faith week. To celebrate that milestone, the affiliate worked on a total of twenty-one houses that week, commemorating our twenty-one years as a building ministry. One of the houses was jointly sponsored by a mosque and a Methodist church. Other houses were sponsored by a wide array of Christian churches and also by a synagogue and a Baha'i group.

Methodists and Latter-day Saints

In Richmond, Virginia, a house was sponsored by a Baptist church and a congregation of Latter-day Saints. On the other side of the country, in Alta Vista, California, a United Methodist church partnered with another Latter-day Saints congregation to sponsor and build a Habitat house.

A Catholic Church, a United Church of Christ, and a Temple

In Toledo, Ohio, Maumee Valley Habitat for Humanity brought three very diverse congregations together for a special build in June 1997. The three congregations were Gesu Catholic Church, Sylvania United Church of Christ, and Temple Shomer Emunim. The house was built in just nine days. The partnership was such a positive experience that the three congregations built another house together in June of 1998 and a third in 1999. Thomas Landberg, executive director of the Maumee affiliate, says that this partnership is "the ultimate expression of the theology of the hammer."

The Theology of the Hammer

This awesome and very wide array of partners in Toledo, Alta Vista, Richmond, and other places is a part of what in Habitat for Humanity is called "the theology of the hammer." What is that? I have written a book by that title, but let me outline the central tenets of the concept below.

First, true religion is more than singing and talking. Action is required. I believe, as a Christian, that salvation comes from God through Christ. Building houses and other good works do not earn me or anyone else salvation. It is God's free gift and cannot be earned, but as an expression of gratitude to God, I should live a life of "giving back." That's done primarily by serving others, doing very simple things like giving water to the thirsty, food to the hungry, clothing for the naked, and shelter to those who are outside. Good work also has the effect of causing people to praise and glorify God. As the Scripture says, "Let your light shine before men, that they may see your good deeds and praise your Father in heaven" (Matt. 5:16 NIV).

The theology of the hammer likewise mandates that, while there is difference in theology, politics, and philosophy, there can be agreement on a hammer. We can lay aside theological, political, and philosophical differences and build together to accomplish our goal of eliminating poverty housing and homelessness from the face of the earth.

Former President Jimmy Carter escorts Lady Bird Johnson to see houses at the completion of the Jimmy Carter work project in Houston, Texas, in 1998. Photo by Robert Baker.

Democrat and Republican

In 1984, I recruited former President Jimmy Carter to work with us in Habitat for Humanity. He started out by helping renovate a six-story building on the lower east side of Manhattan in New York City. Tremendous publicity went out all across the nation as a consequence of his action. He also agreed to mail out fund-raising letters for us.

These two activities on his part resulted in letters pouring into our office both praising us for getting President Carter involved in the work and blasting us for his involvement. The negative letters were along these lines:

"I no longer want to support you because of Jimmy Carter's involvement. I thought you were a Christian ministry. Now you've become an instrument of the Democratic Party. Take me off your mailing list!"

"Jimmy Carter is now with Habitat? He gave away the Panama Canal. I will not give you another dime!"

"Why did you allow Jimmy Carter into Habitat for Humanity? He let all those refugees into our country. And he caused double-digit inflation. You can forget about any further support from me!"

To all these verbal assaults, we patiently responded. For example, I carefully and thoughtfully pointed out that Jimmy Carter was involved in Habitat for Humanity as a Christian and humanitarian and not a Democrat. And often I would get a reply, "Where are the Republicans?"

Since I have long believed in solving problems and not fighting them, I went looking for a prominent Republican. And I found one in the person of former President Gerald Ford. He agreed to see me at his home in Rancho Mirage, California. I asked him to be a member of our board of advisers and he accepted. His involvement was widely publicized, and thus our problem about no prominent Republican involvement ended.

Former House Speaker Newt Gingrich works on
The House that Congress Built in
Washington, D.C., in 1998.

Since then many other well-known political leaders of both parties in the United States have been significant participants in our work. Newt Gingrich, while Speaker of the House, became involved with Habitat. He raised money for a house in his home district in Georgia and helped build it. He served food at a Habitat house built by women. He faithfully wore a Habitat lapel pin. And working with Congressman Jerry Lewis and other congressional leaders, he organized "the Houses that Congress Built" program that involved nearly 400 members of the U.S. House of Representatives out of 435 in building a Habitat house in their home districts throughout the country in 1998. Speaker Gingrich even led a group of congressmen to Belfast, Northern Ireland, in August 1998 to help build Habitat houses there. The U.S. Ambassador to the Court of Saint James's, Philip Lader, also worked one day in late 1998 at a Habitat building site in Belfast. The following day, he and he wife, Linda, worked on another building site in London where four Habitat houses were being built.

Jack Kemp, former secretary of housing and urban development and a Republican vice presidential candidate, served for a time on the Habitat for Humanity International board of directors and, beginning in 1998, served as chairman of our Rebuilding Our Communities campaign to raise an additional $500 million for our work worldwide. In that role, he personally contributed thousands of dollars and continues to guide the campaign to achieve the ambitious goal.

President Bill Clinton and First Lady Hillary Rodham Clinton have long supported the work with both their personal involvement and financial support. The two of them and their daughter, Chelsea, worked on a Habitat house in Atlanta in 1992, during the presidential campaign, along with former President Carter and Al and Tipper Gore and their children.

President Clinton visited a Habitat homeowner in Kanşane, Botswana, in early 1998 during a trip to Africa. A year earlier, Hillary worked with Rosalynn Carter and several other state first ladies and former first ladies in Pikeville, Kentucky, to help build the first First Ladies House. In the months following, several other first ladies from several states also built First Ladies Houses. And plans are being drawn up for the first lady of every state—or the governor, if a woman—to eventually build a Habitat house in each state capital throughout the United States.

Hope Taft, the first lady of Ohio, is very excited about the First Ladies build program. She is working with Terri Bate, executive director of the Canton affiliate, to build a House of Hope in Canton and possibly a Circle of Hope all around the state.

On his fiftieth birthday, Vice President Al Gore worked on a Habitat house in the Anacostia neighborhood of Washington, D.C. President Clinton spoke at a Habitat house dedication service in another neighborhood in the capital.

Vice President Al Gore and President Bill Clinton
hammered on a Habitat project.

Across Borders

Local and state politicians of both major parties in the United States and major political figures in other countries have likewise joined Habitat for Humanity. For example, Ed Schreyer, former premier of Manitoba and former governor-general of Canada, has since 1994 led an Ed Schreyer blitz-build in various locations across Canada. His support has given great visibility to the growing work of Habitat in that country.

As reported in chapter 2, the Honorable Benjamin Mkapa, president of Tanzania, spoke at a Habitat dedication service in Kasula in northeast Tanzania in late 1997. His royal highness the duke of Gloucester is the official patron of Habitat for Humanity of Great Britain. He frequently participates in Habitat events in the United Kingdom.

When Linda and I visited Korea in July 1998, I had a meeting with President Kim Dae-jung. He signed a membership form during our meeting, officially joining the Habitat for Humanity organization in Korea. A week later, in Manila, the Philippines, I met with the new president, Joseph Estrada. He was especially enthusiastic about Habitat for Humanity. When Jimmy Carter was in the Philippines in March 1999 for the two hundred and ninety-three 6house blitz-build, President Estrada came out to work and gave his full support.

Habitat's Church Backbone

So political leaders of various parties and with differing political philosophies are supporting this ministry of hammering out faith and love, but the backbone of our support comes from the churches. And as mentioned earlier, in spite of theological differences, the churches are coming together in unprecedented ways to build houses.

Many churches within a particular denomination often come together to build. For example, in the summer of 1998, seven Presbyterian churches from west Los Angeles partnered to build two Habitat houses with Los Angeles Habitat for Humanity. James Adams, chairman of that coalition of churches, was very excited about that successful venture. He exclaimed, "I was so proud to get seven Presbyterian churches together in one room!" When this book went to press in mid-1999, James and his coworkers had raised their sights and were expanding the coalition to include even more sister Presbyterian churches to build even more houses with Los Angeles Habitat later this year and in coming years.

In Texas, the Episcopal diocese of West Texas started supporting local Habitat affiliates in 1990. In 1994, the annual council of the diocese approved a partnership with Habitat for Humanity to furnish volunteers and money to work with local affiliates. The partnership is called Building Homes and Hope.

By early 1999, five houses had been sponsored and built by the partnership. And more were planned for the months ahead. The Reverend James Edward Folts, bishop of the diocese, is very involved in the venture and totally supportive of it.

In North Carolina, dedicated Habitat partner Dr. Henry T. Clark Jr. of Chapel Hill challenged several parishes in the Episcopal Diocese of North Carolina to build Habitat houses with their local affiliates. He put up matching money to stimulate the churches. As a result of his initiative, Trinity (Mount Airy) built and dedicated a house. All Saints (Roanoke Rapids) broke ground for their Habitat house on December 13, 1998. St. Paul's (Winston-Salem) signed on to build in 1999. Grace (Lexington) is in process of raising their portion of the money for a house as this book goes to press. Galloway Memorial Chapel (Elkin) has agreed in principle to participate in the program. Dr. Clark wants to vastly expand the whole concept throughout North Carolina and beyond.

In 1997, Greeley, Colorado, Lutherans sponsored a Habitat house that was built during Building on Faith week in September. Perry Sukstorf, project leader of the house, wrote about the experience. "Many non-Lutherans wonder why Lutherans are split into three different church bodies, namely, Lutheran Church Missouri Synod (LCMS), the Evangelical Lutheran Church in America (ELCA), and the Wisconsin Evangelical Lutheran Synod (WELS). Many of the reasons come from heritage and traditions while other reasons have to do with differences of opinion about theology," he explained. "This diversity has led many of our congregations to look for ways to reach out in unity to serve Christ in deed, while at the same time, retaining their separate but similar beliefs. Habitat for Humanity is the way our Lutheran community decided to work together on a project that we all could agree served Christ."

Aid Association for Lutherans, a fraternal insurance society headquartered in Appleton, Wisconsin, serves all Lutherans and has been a generous partner with Habitat for Humanity all across the country. This organization gave a twenty-thousand-dollar challenge grant that was matched by the local congregations.

I was privileged to be present when the house they sponsored in Greeley was dedicated at the end of Building on Faith week, along with two additional houses that were sponsored by churches of other denominations.

Eva Ruth, a board member of Fayetteville Area (North Carolina) Habitat for Humanity, wrote about an exciting partnership of twenty-seven Presbyterian churches that came together to build a house with that affiliate:

> In April (1997), our Habitat board voted to build a house in September in connection with our tenth anniversary and Building on Faith week. We brainstormed for several minutes but no real plans were made. No formal responsibilities were assigned. Then in mid-July, the building supervisor realized that time was closing in on their house. There was much to be done in a few short weeks. It was decided to contact the Presbyterian churches with a proposal asking, Would they be willing to organize their people to build this house? The group of Presbyterian churches consisted of twenty-seven congregations scattered throughout the county. They had not met as a group for eighteen months. At first, the leaders balked. No way, they said. But they did agree to send out feelers to see if there was any interest.
>
> In less than a week, a building coordinator had volunteered. Notices were sent out in church bulletins and workers immediately

sent them in. There were even volunteers for crew chiefs. At the same time this was happening with the Presbyterians, the volunteer coordinator sent out requests to other churches in our covenant church program. There were many willing workers everywhere it seemed. A core group met to plan the work schedule.

In the meantime, I was asked to coordinate food for the many workers for the blitz. *No one will agree to feed this many workers in such a short time,* I thought. After praying for direction, letters were sent to all twenty-seven churches. Instead of assigning days, I decided to let the churches call me with whatever help they could offer.

The phone calls began almost as soon as the letters were sent. We needed two snacks and lunches for forty people with at least one volunteer to stay to serve for eleven days. It was as though I was in a dream—no two churches asked for the same day or wanted to bring the same meal!

Our amazing story doesn't end there. The work on the house began. On some days extra food would show up from other churches not included on the schedule. On those days, we almost always had more workers than food. Without the extra food, we would have run out. It was as though the loaves and fish had become our own story.

This house became an ecumenical project with workers from many churches of different denominations. We were from a variety of economic backgrounds and varied skills. Our goal was the same and our hearts became as one as we met early in the morning for worship before beginning our work. It wasn't coincidence; it was God who planned our activities for Building on Faith week. How evident that was throughout our time together. This house was completed in two weeks without any difficulty—a great first for the Presbyterians!

The Saints House

Another exciting and humorous story happened during Building on Faith week in 1997 in Milwaukee. As I was visiting the various houses on the construction site, I came to one with a big sign out front that proclaimed, "The Saints House."

"What does that mean?" I asked Jean Lesher, director of Milwaukee Habitat.

"Well, she replied, "The various Catholic churches in the city with 'Saint' in their name came together to sponsor this house."

"Wonderful," I exclaimed. "I want to go inside and meet some of the 'saints.'"

Walking into the house, I started introducing myself to the volunteer workers, and to my surprise most of them were Baptists!

"How is it that you Baptists are working on the Saints House?" I asked.

"It's simple," they responded. "We thought this was our best opportunity to be saints!"

Lines Crossed

There have been other wonderful examples of people from different churches working together to build houses. All sorts of lines are constantly being crossed in this ministry to the enrichment of all concerned.

One place where lines have been crossed in a remarkable way is South Bend and Mishawaka, Indiana. There the local affiliate, St. Joseph County Habitat for Humanity, has brought together an amazing array of churches, both Catholic and Protestant, to build houses. The churches that have helped build the houses are St. Matthews Catholic, Little Flower Catholic, St. Jude Catholic, St. Bavos Catholic, Holy Family Catholic, Kern Road Mennonite, Crest Manor Church of the Brethren, First Baptist, Grace United Methodist, Lakeville United Methodist, South Side United Methodist, First Presbyterian, Sunnyside Presbyterian, Trinity Evangelical Free, Redeemer Missionary, First Brethren, Episcopal Cathedral, Hilltop Lutheran, St. Peter's Lutheran, and St. Paul's Lutheran.

Former Habitat International board member LeRoy Troyer, who is a leading architect in St. Joseph County and an active member of the local Habitat affiliate, wrote the following about the impact of Habitat for Humanity on the churches:

> Before Habitat for Humanity, church leaders and lay people would seldom meet and did not know each other. Now, after building Habitat houses together with the homeowner families, we are getting to know each other and find ourselves encouraging one another. I find that we are supporting each other in applying our Christian faith in business, in daily life, and in all that we do.
>
> Catholics and Protestants are working together in a good way, respecting each other. Ten years ago it would not have occurred to people that it would be possible. We are learning that the application

of Jesus' teachings is breaking down barriers—all through the ministry of Habitat for Humanity.

Habitat is uniting the faith community not only by providing decent housing for people, but also by bringing awareness of other needs of the economically poor people to the larger community. Two examples come to mind. One, Kern Road Mennonite Church members are volunteering two evenings a week tutoring at-risk children at St. Paul's Baptist Church for neighborhood children. Two, when there is major violence in the area, pastors and laypeople get together with the mayor for prayer and vigilance at the crime location in support of and to encourage people in the neighborhood.

I believe Habitat is building partnerships among the churches that work together to build houses and address other community needs.

Ireland's Catholics and Protestants Together

Of course the most dramatic example of Catholics and Protestants working together in the ministry of Habitat for Humanity is in Belfast, Northern Ireland. The work in Belfast started in 1994. Peter Farquharson, a dynamic Presbyterian layman, is director of the Belfast affiliate. His construction supervisor is Gerry Crossin, a Catholic and former member of the IRA.

Initially, the building was done in Iris Close, a Catholic neighborhood. By 1998, eleven units had been completed. As mentioned earlier, Newt Gingrich and other political figures worked there in August 1998. Then work started in Glencairn Estate, a Protestant community. By early 1999, six homes had been built there. A total of sixteen are planned.

Belfast Habitat for Humanity hopes eventually to build an integrated project—Catholics and Protestants living in the same neighborhood—to model the Good Friday (1998) peace accord. Peter Farquharson says about the work in Belfast, "The poverty we face in Belfast and Northern Ireland is not just physical but spiritual. We need genuine reconciliation. At Habitat, our mission is clear: to call the church into action, to build houses, and to rebuild community."

An Ex-IRA Member's Reconciliation

A wonderful example of reconciliation is that of Gerry Crossin, the construction supervisor mentioned above. I met this remarkable man when I visited Belfast with Washington, D.C., Habitat International director Tom Jones in 1996. Jerry spoke one evening in a Protestant church. He told about his life with the IRA and about some of his activities with them. He shared quite openly how he had a seething hatred of Protestants. Then something happened. God came to him one night, he said, and lifted the hatred from his heart. It was wonderful, he exclaimed. He went on to say that he is now able to love Protestants as much as he loves Catholics. "I like to build a house for a Protestant family," he said amidst tears, "as much as I like build for a Catholic family." Everyone in the church that night was incredibly moved by his powerful testimony.

Australia, Canada, Honduras

In Australia, there's a wonderful story of a different kind of coming together of the churches. In the city of Whitehorse, which is a part of Greater Melbourne, there are eighty-three churches of a dozen different denominations. Sixteen of those churches joined forces to form an affiliate of Habitat for Humanity. Members of those churches serve on the board of management, various committees, and as volunteer builders. Dr. Howard Fearn-Wannan, president of Habitat for Humanity of Australia, says, "Habitat for Humanity could well become a great uniting force in the badly fragmented Australian church."

In Canada, there is division and tension between the dominant English-speaking population and the French-speaking Canadians, who are largely concentrated in the province of Quebec. On September 15, 1997, the first day of that year's Building on Faith week, Habitat for Humanity of Canada began building its first house in Quebec, in the city of Aylmer, which is right across the border from the national capital of Ottawa.

I was privileged to be present on that historic occasion. That evening, a local United Church of Canada was filled with worshipers offering thanks and praise to God for the beacon of reconciliation in their country. I spoke in both English and French, sharing the Habitat vision. At the end of the service, seven clergymen from various churches—from a Roman Catholic archbishop to a pastor of a local Bible church—offered prayers of blessing and benediction.

At the end of that same week, I was in San Pedro Sula, Honduras, where a blitz-build of twenty-one houses was concluding. Scores of churches of many denominations were actively involved. Melvin Flores, national director of Habitat for Humanity in Honduras, wrote about how they got the churches on board.

"In San Pedro Sula there are about two hundred and fifty churches including Evangelical, Catholics, Mormons, Jehovah Witnesses, Episcopalians, Adventists, and Orthodox. We knew that our biggest challenge was obtaining the participation of the local churches. So we asked God to open the doors for us, for His grace to touch especially the hearts of the shepherds and leaders of the various congregations," Melvin said.

"We started first by establishing contact with more prominent people in the community and asking for their advice on how to stop the churches from talking only of what is spiritual and helping them understand that the spirit is inside a physical body and this body needs food, clothing, and a decent roof to live under," he explained. "This was how we approached the association of evangelical pastors of the city. After our presentation to the association, we proceeded to visit each of the churches. We also visited the Catholic church and the Episcopal church. They were more open to this type of activity, which they called 'social projection,' and straightaway they committed themselves to support us."

They received several promises from various churches, he explained, but they had to wait until the actual date for Building on Faith week to begin to see if they really would fulfill their promises. "For them, like for us, it would be a first-time experience, and we could not be sure of the results."

Finally, the big day arrived. More than two hundred people were present for the first day of work. Throughout the week, they averaged around one hundred and fifty volunteers from the churches every day. "During the week, religious and doctrinal differences were set aside," he reported happily. "We experienced wonderful things and we got the houses built."

Building on Faith week was concluded that night in a local church with a traditional Habitation service. I spoke, but the most important words were those of a young twenty-four-year old woman named Syrian, who had helped build that week. Here's what she said:

> I walked on the earth for what seemed a long time without finding something that could really give sense to my life. Then I accepted Christ in my heart. And my life began to change. Still I felt that something was missing. Today I give thanks to God and to Habitat for Humanity because it was in participating in the event of Building on Faith that I witnessed the wonderful experience of partnership and integration between the young people of my church and the homeowner family. I wish that many more institutions like these existed, allowing a true partnership between people of all ages, education,

philosophy, and means. Today, after such an unforgettable week of work with the Habitat families, for the first time I feel like a true Christian, not only in words, but also in deeds. I now understand that all of us can serve our neighbors in need.

Millard Fuller is seated next to an Ashanti chief at the celebration of Habitat for Ghana's one thousandth house, which was built in Wamfie. The African attire was a gift presented to Millard during the celebration in January 1998.

Crossing Lines in Africa

In Ethiopia, Habitat for Humanity is bringing the churches and other religious bodies together. Wubetu Tadesse, president of Habitat for Humanity in Soddo, a town in the south of the country, shares about how Habitat is crossing all lines and bringing people together in his city. I met Wubetu in 1995 when Linda and I visited there with Africa director, Harry Goodall. At that time, Habitat was just getting started in Soddo. Wubetu's wife had recently had a baby boy, and since Wubetu was in the capital city of Addis Abbaba at the time of the birth on Habitat business, the little boy was named Habitat. We met and held the wee fellow. Our picture with baby Habitat is in my book, *A Simple, Decent Place to Live.*

"Habitat for Humanity's clear idea brings different religious bodies to work together," Wubetu said. "The Catholic church promised to cover one-third of the cost of the water-line installation to the building site. The

Protestants promised to help with construction material transport services and to also help with water line installation. The Muslim leader in Soddo exclaimed that he appreciates the mission focus of Habitat for Humanity. He agreed that the theology of the hammer is powerful and a true idea."

In Zambia, Habitat for Humanity is also bringing the churches together, but it is doing more than that. Waveson Hamani Hamuchankwi, a Habitat leader in the Chisamba affiliate, near Lusaka, explained how Habitat is bringing churches and people of different economic levels together in his part of Africa:

> In Chisamba, different church groups do not want to associate with other church groups because they think those others do not worship a true God. Some are Catholics and some are Protestants. Also people are from different backgrounds. Some are rich and some are poor. These groups find it difficult to mix together because rich people think poor people are lazy or they like being poor. Poor people do not like rich people because they think the rich do not want to share what they have with others who are poor. We told the people that Habitat for Humanity was for the poor people with the support of the rich people who feel they should help change the lives of their poor neighbors regardless of their religious affiliation, race, tribe, or any ethnic differences.
>
> People were amazed. They did not understand how we were going to work because they did not see themselves working with the rich or the poor and neither could they envision Catholics working with Protestants or Protestants working with Catholics.
>
> We started our work by educating people first on the plan of God to see man exist peacefully and help each other. When we saw that they understood us, we asked communities to form committees. From there the people were able to know that it is possible to work together in spite of coming from different backgrounds.
>
> At our Habitat functions we include all singing groups from various churches in the area. The poor and the rich now mix freely without feeling bad about the other group. Habitat for Humanity has managed to bring these different groups together.

Crossing Lines in Asia

In Korea, Habitat for Humanity is bringing the churches together. Linda and I were there in July 1998, as mentioned earlier, along with Asia and

South Pacific director, Steve Weir. On a Sunday morning I preached at the Kum Ran Church, the largest Methodist congregation in the world, with eighty-five thousand members. That afternoon I preached at a Presbyterian church in Ui-Bong-ju. Both churches support Habitat in Korea along with many other churches of various denominations.

On that same trip, we were in Malaysia, where Habitat was just getting started in the city of Kuching. I preached that Sunday in St. Faith's Anglican Church and First Baptist Church. Both of these congregations and several others were organizing Habitat for Humanity in that city. John Chin, pastor of First Baptist, is president of the new Habitat affiliate.

Raising Awareness

Sometimes churches organize special events to raise awareness and funds for Habitat for Humanity. In so doing, the participating congregations are drawn together.

One such event was a Two-by-Four Sunday in Rockwell, Texas. The cooperative effort was organized by First United Methodist Church and Our Lady of the Lake Catholic Church. The event raised seventy-seven hundred dollars, enough to pay for a roof, bricks, and rebar for the foundation of the seventh house of the local Habitat affiliate.

Six congregations in South Bend, Washington, came together for a special purpose which resulted in helping the local Habitat affiliate, Habitat for Humanity of Willapa Harbor, build their first home. The special purpose was a community Vacation Bible School and the participating churches were from three different groups—United Methodist, Lutheran, and Catholic.

The theme of the Bible school was Be a Promise Builder for Jesus. Habitat for Humanity was selected to receive the offering. Each day the eighty children learned about Habitat and brought their offering. By the end of the Bible school, the children had given nearly three hundred dollars for that first house.

An Interfaith Calling Tree

Jeanne Brokshire of Habitat for Humanity, Mat-su in Wasilla, Alaska, tells of a very unique and powerful way churches in her area helped their work:

We had a very serious problem. We wanted to move in the right direction but didn't know which way to go. The new board of directors prayed for a solution and also put out a call for prayer throughout the valley. Within four hours of the request for prayer, hundreds of people that make up the prayer chains of twenty-four churches in the Matanuska Susitna Valley in Wasilla and Palmer were petitioning God for answers.

A representative of the prayer chain from each of the churches was put on a "calling tree." Dr. Milt Lum, a member of the board, called the first three: a Lutheran, a Presbyterian, and an Episcopalian. Each of these prayer-chain representatives started the prayer around their own chain and called two other church representatives who repeated the process until all the participating church prayer chains were in operation. This included Catholics, Baptists, Seventh-Day Adventists, Methodists, Nazarenes, Assembly of God, Pentecostals, and many others.

Within days, the board of directors reached a decision, causing a few immediate problems. In the year ahead, though, the decision proved to be extremely beneficial to our affiliate. And all thanks were given to God for His guidance at that time of crisis.

When the Christian community, regardless of denominational differences, unites to petition God for solutions, He hears. Besides informing God's people that Habitat was prayerfully seeking answers for tough decisions, the church community built bridges among themselves. And in the months ahead, the churches worked together to build the first Habitat house in Wasilla, Alaska.

From Five Denominations

Kim Gabriel, the dynamic executive director of South Brevard Habitat for Humanity in Melbourne, Florida, tells the exciting story of six churches of five different denominations that worked together to build a house with and for a single mom and her two young children. Kim says that not only was the house put up, but so much more was built too.

"During the dedication of the completed house," Kim reported, "volunteers from all six churches held hands and made a circle around the house, with each of the six pastors taking a turn at praising God and blessing the house. It was a very moving experience. That evening a 'thank-you' gathering was held at one of the churches for all who had participated in the build. When communion was offered, volunteers from five denominations went to the rail and shared communion. This was a sight to behold and the theology of the hammer at its best."

The churches that participated in the build were St. Mark's United Methodist, Palmdale Presbyterian, First Baptist of Satellite Beach, Church of All Nations, St. John's Episcopal, and Pineda Presbyterian.

Crossing Racial Lines

In Dexter, Georgia, the churches came together to form a branch of Dublin-Laurens Habitat for Humanity. Local businessman Roger Lord was the principal organizer of the group. A total of seven congregations joined the effort—Antioch Baptist, Dexter Assembly of God, Dexter Methodist, Dexter Baptist, Buckhorn Methodist, Mount Carmel Baptist, and New Bethel Baptist. Each church donated fifteen hundred dollars and raised sixty-five hundred dollars in a joint fund-raiser to help pay for the first house.

Meanwhile, in nearby Dublin a coalition of black and white Baptist churches were raising money for a Baptist Habitat house. The participating churches were Dudley Baptist, First African Baptist, Dublin First Baptist, Jefferson Street Baptist, and Springhill Baptist.

Another house that was built by white and African-American congregations was in Richmond, Virginia. And the house was built for a remarkable woman and her four children. "Sudie Williams, her husband, John, and their children moved to Richmond in 1987," explained Mary Evans of Richmond Metropolitan Habitat for Humanity. "They thought they had work lined up in Richmond, but both promised jobs fell through. They had little money and no place to live. They found rooms in a motel and then moved to an emergency shelter. All five members lived in one room while Sudie and her husband looked for work and a place to live.

"The Williams first heard of Habitat for Humanity from another couple at the shelter who were applying for a house with Richmond Habitat. Sudie got an application for a Habitat house but didn't send it in," explained Mary. In the months that followed, the family moved several times into various rundown apartments. During this period, Sudie's husband left to return to his family in Tennessee. She was on her own.

"Sudie found a job at a day-care center. But she was still in need of decent housing. She had applied twice in two years to the Richmond Redevelopment Authority for low-income housing and was told both times she didn't make enough money," Mary continued. "About this time, she found the application for a Habitat house. She filled it out and mailed it. When two members of the family selection committee visited the family, they found asbestos falling from the ceiling, waste water coming from the upstairs apartment bathroom, and heating bills that were astronomical because of holes in the walls."

With high hopes, Sudie and her family began working on other houses for Richmond Habitat. They had completed eighty hours of sweat equity before they were approved for the program. With the help of her advocates, George and Jane Hastings, and a friend who dug the foundations with a bulldozer, the family completed their three hundred and fifty hours on Good Friday in 1994. The walls for their house were raised on that special day. Some of her church family, Gates of Faith Ministries, helped build and provided assistance later.

Two other churches worked on Sudie's house—Metropolitan African American Baptist and Northminister Baptist. Sudie was ecstatic. "It was the greatest experience, being a church person myself," she exclaimed, "to really see God's word put into action."

Dedication day for her new house was October 2, 1994. Mary said it was wonderful; the diversity of the crowd was a sight to see. "The choirs from the two churches joined together for the occasion—black and white blending together to sing of joy and celebration."

Sudie and her family have blossomed since their involvement with Habitat for Humanity. Her children have received scholarships, some from local churches affiliated with Richmond Habitat, and all have now completed college or are headed for college.

Her oldest son, Troy, graduated from Norfolk State University with a degree in business. Peter graduated from Virginia Commonwealth University with a degree in accounting. Daughter Tiya graduated from the University of Virginia with honors in sociology. And the youngest, Paul, is a student at J. Sergeant Reynolds in Richmond, as of when this book went to press in 1999. Sudie became president of the Richmond Habitat homeowners association and became the first president to serve on the Richmond Habitat board of directors.

So the theology of the hammer is bringing all sorts of people together—politicians of different parties and political persuasions, churches of virtually all denominations, liberals and conservatives, rich and poor, people of many races—all to work together to build better lives for people who need a simple, decent place to live.

CHAPTER 13

Pursuing Happiness

Today is the happiest day of my life.
—LEONARD TYSON, HABITAT HOMEOWNER

"W e hold these Truths to be self-evident," states the Declaration of Independence of the United States, "that all Men are created equal, that they are endowed by their Creator with certain unalienable Rights, that among these are Life, Liberty, and the Pursuit of Happiness." The Declaration goes on to affirm that it is the right of the people "to insti- tute new Government, laying its Foundation on such Principles, and orga- nizing its Powers in such Form, as to them shall seem most likely to effect their Safety and Happiness."

So in the United States, since the inception of our nation, it has been a firm principle that the pursuit and realization of happiness is a very desir- able thing. Government should always direct its efforts toward principles and practices that have the effect of producing happiness among the people. By inference, all other organizations within society and individuals them- selves should pursue activities that ultimately produce happiness. There is duty, responsibility, sacrifice, and much more in life, of course, but the end result of all of it should be happiness. As the Declaration of Independence states, our Creator endows people with certain rights, including happiness.

But this pursuit of happiness is not restricted to the United States. Without a doubt, human beings around the world pursue and desire happiness.

What Brings Happiness?

What brings about this happiness? Obviously, there are many factors, and they are as varied and complicated as human beings. That which would bring great happiness to one person might well bring distress and unhappiness to another. Some things, however, seem to be universal in bringing happiness to people.

First, there is love. Every normal person on earth likes to feel loved. And that feeling of love brings happiness. Giving to others out of a heart of compassion and concern brings happiness to the giver. Likewise, the recipient of love is made to feel happy.

Believing that God is love and that God loves us *individually* brings enormous happiness and a contentment that transcends even tragedies in life and death itself. An assurance that God is in control and that His love for us *individually* extends even beyond the grave gives comfort, peace, and happiness.

A sense of well-being, of being in control of things affecting one's life, tends to create happiness. Purpose and meaning in life also bring about happiness. And the well-being of loved ones and friends tends to usher in happiness. Feeling like you are making progress in life, that you are progressing toward a desired goal, tends to bring about happiness.

Many other factors can also bring about happiness, but something that is quite universal in producing happiness is a good and decent place to live. And when that place to live is made possible by a group of people working together out of a love-and-faith motivation, the happiness is multiplied many times.

There are several stories that document the dramatic effect houses have on the happiness of the families that receive them. And I want to tell you about some of the happiest of them. These stories show how this happiness is not restricted to just the family members of the new homeowners, but also to the volunteers, the donors, and the Habitat staff.

We are building more than houses. We are building happiness.

Happiest Day of My Life

Listen to what Shari Bell of Canton, Ohio, has to say about the happiness her Habitat house brought to her:

> I purchased my Habitat home on November 13, 1998. That was the happiest day of my life. God truly blessed me and my family. I have a foundation and stability. My fourteen-year-old son is a young scholar because of Habitat. His college is paid for at Ohio State University for four years because of that solid home foundation. My

seven-year-old is doing well in school, too. I have had so many blessings come from my partnership with Habitat for Humanity. I went to Nicaragua and helped build homes. I have learned to paint, do repairs, and make cement blocks. Most importantly, I have made friends and have been able to spread God's love. I am still a volunteer for Habitat, serving on the family selection committee and as a family advocate. I also speak at fund-raisers. To me, Habitat means God's love can conquer all. Thank God for Habitat for Humanity!

Another family that was made unbelievably happy by their Habitat house was that of Joe and Donna Eller.

When Joe was eighteen, in 1976, he was walking along a highway near a small mountain town in North Carolina. He had been drinking and abusing drugs. In his state of intoxication, he staggered into the path of an oncoming car and was severely injured. Joe lay unconscious for two months. His family was told that if he lived he would be a vegetable. But after several operations over the next few years and then months of physical therapy plus Joe's determination, he regained the use of his legs and one arm and hand.

In 1979, Joe enrolled in a sheltered workshop. Again his determination to succeed made him one of the most valued participants in the workshop. Soon he was hired at a short-order restaurant. His dependability and earnest work ethic won him the Job Training Partnership Act Adult Participant of the Year award. It was presented to him by Governor James Martin. Through the workshop, Joe gained financial stability and a degree of self-sufficiency. He also met Donna, who became his wife. Together they rented a tiny apartment that was owned by the Sheltered Workshop. Then in 1993, with the aid of the director of the workshop, Joe and Donna submitted an application for a Habitat house.

Sondra Edwards of Watauga County Habitat for Humanity, Boone, North Carolina, describes what happened next: "When the family selection committee chairperson went to tell Joe and Donna that they had been selected to have a Habitat house, Joe exclaimed, 'This is the happiest day of my life and do you know what else . . . it's my birthday!' Joe and Donna worked patiently and diligently as their skills would allow and they logged over five hundred hours of sweat equity in their eight hundred and fifty square foot home. They warmly thanked everyone who came to volunteer. There were more than three hundred people who assisted with their house," Sondra reported.

"Finally, on a much-anticipated day in September 1996, Mr. and Mrs. Joe Eller were handed the keys to their new home on 101 Woodpecker Lane. Amid many misty eyes, Joe took the keys and, in his halting yet sincere speech, he related his story and said, 'The doctor said that I wouldn't be able to walk or talk and that I would be a vegetable. Well, you can see that I'm walking and talking and I ain't no vegetable. Thank you for this house. Without Habitat, I couldn't pay the interest much less the principle!'

"Since becoming homeowners, Joe and Donna have learned budgeting skills and have never been late with a house payment," Sharon added. "Donna, also employed at a short-order restaurant, is taking cooking lessons. Both work very hard at keeping the house nice."

Incredibly Happy

Leonard Tyson in Americus, Georgia, is another person who was made incredibly happy by his new Habitat house. He is well known in Americus. Leonard is legally blind although he has limited peripheral vision. He is a talented musician who plays and sings in churches in the area and also performs at other events. But as talented as he is, in the small town of Americus, he could not make enough money to get adequate housing in any conventional way. It came to our attention at Habitat that Leonard was living in a pitiful shack on the north side of town. Not only was the house grossly rundown in terms of roofing, siding, flooring, and so on; it had no water. I'll never forget going out to Leonard's house one afternoon with Ted Swisher, then director of the local Habitat affiliate in Americus.

"Leonard," I asked, "since you have no water in the house, where do you go to the bathroom?"

"Up the street at a service station."

The small gasoline station was two blocks away. I was shocked and deeply saddened as I reflected on this man's situation. Day or night, rain or shine, boiling hot or freezing cold, if he needed to get a drink of water; wash his hands, face, or body; or even use a toilet, Leonard had to walk two blocks and hope that no one else would be using the service station rest room.

Leonard was chosen to have the annual Christmas house that is built each year by Americus-Sumter County Habitat for Humanity. Volunteers and Habitat staff people worked diligently over a four-week period in November and December to build the house. Then, on the Saturday before Christmas in 1995, a large group of family, friends, and Habitat builders gathered in the front yard of Leonard's brand new one-bedroom, one-bathroom, eight-hundred-square foot house for a moving dedication service.

Leonard has a few private students whom he teaches the piano and other musical instruments or to sing. Some of his students performed at the dedication. Others on the program spoke about the build and recognized special people, including individuals and organizations who contributed to the house.

I spoke and wished Leonard a long and happy life in his house. Then a Bible and the keys were presented to him, and he was given an opportunity to speak. Leonard stepped slowly to the microphone.

"I don't know what to say," he intoned. "I am a musician. It doesn't bother me to sing or play in front of large crowds. But I am not a public speaker. I really don't know what to say other than to say that this is the happiest day of my life."

Leonard slowly ambled back to his seat. As I wiped the tears from my eyes, I noticed many others doing the same thing.

In the months and years since Leonard moved into his modest new home, I have often driven past it. Every time I do, a wave of happiness comes over me. I remember Leonard's happiness on that dedication day, and I think about how he now only needs to walk across a room to use the bathroom and not two blocks. Just knowing that Leonard is in that modest, good, solid house fills me with joy and great happiness.

Couldn't Believe the Happiness

Amelia Silguero and her family were chosen to have the second house built by Habitat for Humanity–Corpus Christi, Texas. She tells about her experience and about the happiness that has come to her by receiving a Habitat house:

> My association with Habitat started in 1990. I had two co-workers who served on the family selection committee for Habitat for Humanity-Corpus Christi. They thought I could qualify for a Habitat house. When they realized that I would, they approached me with the idea, and I felt it was too good to be true. I had always wanted my own home but felt it was a dream and could never become a reality. With the help of my co-workers, I filled out the paperwork, went through the interviews, and awaited the result of the selection process without believing that something so good could happen.

Then one day my co-workers said that they had good news. They told me I had been selected to receive a Habitat house. They told me about the sweat-equity hours that were required of me and my family during the construction phase. I just said, "Good," and went back to work. A few hours later, I asked one of my co-workers, "Did what you told me mean what I think it means? Am I *really* going to get a house?" When it was confirmed that I was indeed going to get a house, it really sank in! I felt weak and overcome with excitement. I couldn't believe it! We were going to have a home!

The process was long. I was promised several completion and move-in dates that came and went. However, in early 1992, we finally had the dedication, and I moved in the following week. The dedication was special because my family was all there and even the bishop of the diocese of Corpus Christi of the Catholic church. My co-workers, Barbara Gilbertson and Cindy Zipprian, presented a Bible and a crucifix to our family.

The next day a huge picture of me and my family, with me in tears, appeared on the front page of the *Corpus Christi Caller-Times* newspaper. It really had to be the happiest day of my life. The tears I shed were tears of joy.

Since moving in, I have come to realize how nice it is to have my own home. I no longer have to call and wait for landlords for repairs. I have lived in some houses where opossums came in and ate my bread that was wrapped up on my dining table. I have lived in houses that had no working plumbing and landlords who would not make necessary repairs. Now I have a home of which I am in control. If something breaks, either I or someone in my family can repair it. Even though I have to pay for the repairs, the house is mine. In several years, the house will be paid off and I can enjoy the fruits of the labor that we put into getting the house ready. I thank God for the opportunity to associate with Habitat for Humanity!

Happiness Seen in the Children

The happiness of the Garcia family over getting a Habitat house was most dramatically shown by their children. Joe and Angie Garcia were living in a tiny house with one bedroom in Grand Island, Nebraska. Angie said that the house had very little insulation and their gas bills were sky high. The wiring was defective, the plumbing was bad, and the carpeting was old. Angie had to continually scrub the aging carpet whenever the kids spilled something. The kitchen was so small the family had to eat in the living room on TV trays.

The Garcias learned about Habitat for Humanity and applied for a house. The family selection committee of the local affiliate, Grand Island Habitat for Humanity, visited them and all agreed that the family needed a new and bigger house.

In November, Angie got a call from the Habitat office. "Angie, are you sitting down?" asked the voice on the other end. "You were chosen!"

"I couldn't believe it," Angie recalled. "The kids were screaming and jumping up and down. Jessica already had ideas about how she would do her own room."

The family joyfully put in hundreds of hours of sweat equity. The work went along rapidly and very soon the family was ready to move in.

When asked to share the most exciting part of the whole experience, Angie responded, "All this time and effort from everyone was just for us. Nowadays, people are just doing their own things for themselves. But these people give all they could to help other families who needed a home."

Happiness from Togetherness

The Sprunger family in Houghton Lake, Michigan, is thrilled with their new house. Their happiness derives from the knowledge of the motivation behind the whole process.

"As we look back on the events that have taken place, starting on June 27, 1997, we see a group of businesses and volunteers working hard together to make a dream come true for a family they didn't even know," they wrote. "Side by side we worked together, ate together, laughed together, and prayed together. It is all of this that makes this home even more special to us. One can hire a house to be built, but not everyone can say that their home was built with both pride and love like we can. There are no words that can express the happiness we feel. With all the love that went into building our home, we think it is wonderful. We are grateful for our home because it was filled with love even before we moved in."

Another family in Michigan, the Nickersons of Hastings, was equally happy to get a Habitat house. In fact, Amy Nickerson was so happy when she learned their family had been selected to get a house, she ran outside screaming and telling everybody, "We got it! We got it!"

Scott and Amy and their children were living in a tent at one point. Then they moved into a small two-bedroom rental house. Their new Habitat house has three bedrooms and is specially designed for their daughter, Tabatha, who has a gene disorder. Doorways are wider than normal and the house has a ramp for getting Tabatha in and out of the house easily. Light switches are also designed to be easily turned on and off for the benefit of Tabatha. The Nickerson house, built by Barry County Habitat for Humanity, was dedicated on August 23, 1998, with great joy, celebration, and happiness.

A Mysterious Phone Call

Connie Young of Middletown, Ohio, became involved with Habitat for Humanity through a mysterious phone call that led her to tremendous happiness. Later, Connie described her experience to me:

One spring evening in 1995, my phone rang and a lady asked, "Is this Connie Young?" When I said yes, she replied that she had heard that I wanted to volunteer for Habitat for Humanity and that she had something in mind for me. I told her that I was not familiar with Habitat for Humanity and had not offered to volunteer. I did say that I would like to hear more about the organization. The lady told me that Habitat builds homes for people who would otherwise not be able to afford to buy their own home. I told her I would be interested in helping out through my church, but it sounded to me that I was more likely to qualify for a home. She then told me to contact Middletown Habitat for Humanity for an application for a Habitat house. By the way, it was later determined that the lady was calling from a town near Cincinnati and that she had no idea she was calling Middletown, or from where she obtained my name and phone number.

I did call Middletown Habitat and left my name and address on the answering machine. Within two days, I received an application. I put it on my kitchen table and it sat there for about three weeks until I decided to fill it out and mail it back to the Habitat office.

Within two days there was knock at my door, two lovely people were there from the family selection committee to interview me and my son, Philip. By the next Habitat for Humanity board meeting, we were selected to receive a home. That was truly one of the happiest moments of my life.

Working on my sweat equity was hard, but it was a "good" hard. My mother tended to my son so I could work every Saturday and any other free time there was. Working full time, working on my hours, and being a single mom was tough. If it had not been for my mother's help, my church's help, and God's help, I couldn't have done it.

In June 1999 we will have been in our house four years. Habitat for Humanity has become my mission in life, and I have worked with Habitat ever since I first heard about the ministry. I couldn't be more proud of my home. My son and I wake up every morning thanking God for all His blessings through Habitat for Humanity. Each dedication I attend, each banquet at which I speak, only solidifies my gratitude for all the work that is put into this organization every day.

Happiness Galore

Joy unspeakable and happiness galore is what came to Jennifer Kincy in Boca Raton, Florida. Her story is one of those "Believe It or Not" kind. Daria DiGiovanni of Habitat for Humanity of Boca Raton–Delray Beach recounted the incredible series of events that led to the tremendous happiness of this young woman.

"For Jennifer Kincy, there is no truth to the saying, 'You can't go home again.' Thanks to her determination, hard work, and involvement with Habitat for Humanity, she has done exactly that . . . and the result is nothing short of a miracle," said Daria. "Jennifer was born in Gaston, Florida, and moved to Boca Raton with her family at the age of two. When she was eight years old, she went to live with her grandmother along with her mother and siblings. The eldest of five children, Jennifer learned early about responsibility. Although this was a difficult challenge for a young girl, Jennifer now realizes that the lessons she was taught have helped her in raising her own children, Lloyd and Darrian.

"Her grandmother played a major role in Jennifer's development from childhood to independent young woman. When she was twenty-two, Jennifer's relationship with her grandmother took a bittersweet turn. As a result of her grandmother's diagnosis with cancer, the bond between the two women deepened," Daria explained. "The year that followed was the most difficult time of Jennifer's life. She spent most of it in denial until the evening of her grandmother's death."

Years later, Jennifer borrowed money to try to save her grandmother's beloved house, which she discovered at the last minute, was behind in taxes.

Her efforts notwithstanding, the house was condemned and eventually knocked down. Later, when Jennifer applied to Habitat for Humanity for a home for herself and her two boys, her only wish was to get a lot in Boca Raton, a feat considered nearly impossible, given the lack of available lots in the city. Yet even during the application process, Jennifer always believed she would be approved for a home, and she continued to put in her sweat-equity hours on weekends. But nothing could have prepared her for the fateful phone call she received on a warm summer day in mid-August 1996.

"When they called to tell us they had a lot and gave me the address, I just started screaming and dropped the phone," Jennifer said.

The address was her grandmother's lot.

"I had chills up and down my spine," Jennifer added. "It was like a blessing from God. I believe my grandmother had a part in it also."

"Jennifer says that it felt like home for her and the boys from the moment they moved in," reported Daria. "From the neighbors to the mango trees, everything is the same. Perhaps with one difference: Her grandmother is watching over them from a higher place."

Great Golden Hearts

Up the coast from Boca Raton, Rosalinda C. Ponce-Gabor and her children found great happiness in their new Habitat house in Vero Beach. She conveyed it in a letter to Indian River County Habitat for Humanity, the affiliate that built her house.

> I would like to express our heartfelt gratitude to you for providing a house for my family.
>
> Every day I thank the Lord for guiding me to this community, which was unknown to me. Somehow God gave us a solid ground with a friendly, loving, compassionate, caring, and awesome people.
>
> I do not have my parents, brothers and sisters, and other relatives in this country but God provided us an extended family here in Vero Beach, people with *great golden hearts*. It is such a joy to find people like you, who accept us just like a family.
>
> It is an honor and a very satisfying experience to have this home. My gratitude to the Lord and to all of you who have helped me cannot be measured. You have so many things to do in your life, but you chose to help us. I thank all of you for finding time to help needy people like us.

It is just like a beautiful dream that I do not want to wake up. Before, we did not have a safe place, only agony and fear. But now, in this house, we have peace! God is always the answer, and this house is the greatest breakthrough in our lives.

We hope that one day my children and I will be a great influence in this community.

Thank you very much.

Happiness Around the World

In other countries the story is much the same. Habitat houses are bringing great happiness to people.

Daniel and Ann Marie Pilot and their three children moved into their new Habitat home in St. Helens Park, New South Wales, Australia, in 1997. Previously, they had been paying very high rent for an inadequate apartment. Elizabeth MacPherson of Habitat for Humanity, New South Wales, the affiliate that built the house with the Pilot family, described how they feel about their new Habitat house: "The family is 'over the moon' with their new home and are really appreciative to the Habitat people for their support and also for the way they were made to feel special. Daniel works as a bricklayer, and Ann Marie does part-time work to help with finances. The family is very happy with their home and stated that if it had not been for Habitat for Humanity, their lives would have been a lot different as they would still be paying high rent, which would have been very difficult for them. Daniel and Ann Marie enjoyed being part of the team that built their house and are looking forward to working on other homes."

In Poland, Habitat for Humanity is building houses in the city of Gliwice. One of the new homeowners, Joanna Sojka, who moved with her family into their new Habitat house in 1997, writes about the excitement and happiness the house has brought to them:

The day we entered our new home was one of the most beautiful days of our life together, and we will surely treasure it in our memories for a long time. Such a radical change in our housing still has a great impact on our family. For nine years, we lived squeezed into thirty-eight square meters—five people in two tiny rooms with an even smaller kitchen. It's not any wonder that such living conditions influenced us in a negative way—parents, as well as children.

It is hard to believe we lived a normal life in our former housing situation. It was the hope connected with our anticipated move to our new Habitat house that helped us endure a really difficult time.

Today, what had been just a quiet dream for years is fact. We are in our beautiful house now, and our family is waking to a completely different, rather more normal life.

For the first few days in our home, the children ran up and down the stairs as if they had gone berserk, because of all that space they had been given. We now have the privilege of having the entire family over for the holidays. Even an after-dinner nap or work on a hobby takes on new meaning. The one certain fact among all of these is that our dreams have become reality thanks to Habitat for Humanity.

Millard Fuller speaks and leads cheers at a celebration in Chisumbu, Zambia, where Bibles and keys were presented to new Habitat homeowners. Photo by Linda Fuller.

In South Africa, Habitat houses are being built in several areas, including a place just north of Johannesburg called Orange Farm. The affiliate building there is called Habitat for Humanity Arekopenang. Bessy Netsianda of that affiliate wrote about a determined lady who worked diligently to get her Habitat house. She described how this woman's dedicated and determined efforts won over her critics, including her husband, and how her actions brought hope and happiness not only to her own family, but to many others as well.

"Mrs. Serebo was informed that in order to qualify for a Habitat house, she must help with fund-raising. So she started selling meat and vegetables by going around the community with her items for sale in buckets," said Bessy. "Her neighbors discouraged her by saying that it was too hot to sell meat and vegetables from buckets. It was not only the neighbors; her husband was also discouraging her. She was determined to prove them wrong.

"When she finished her fund-raising period, she paid her deposit on the Habitat house. This was when her husband started coming with her to the meetings, because he could see that she was about to get a house. We have recently built her a house and all the neighbors are admiring her, saying that they never thought that selling meat and vegetables in small buckets would lead to obtaining a house," Bessy continued. "She is now recruiting people in the community to come and have houses built by Habitat for Humanity. Her husband is very proud of his wife, as he had given up on acquiring decent accommodations. Mrs. Serebo feels fulfilled to be in a proper house and praises the Almighty for the courage and determination to give her strength to build her house. God is great and may the spirit of the Lord be with the partners of Habitat who helped the families fight the disease that is in the shacks. Mrs. Serebo says that she wishes there could be no more shacks in Orange Farm."

In India, Habitat for Humanity has been building houses since 1985. More than five thousand houses had been completed by early 1999. One place where Habitat is building is in Kanyakumari, the second smallest district in Tamil Nadu, in the far south of the country. The Reverend Retnaraj, president of that Habitat group, wrote about a poor widow whose life was vastly changed for the better by getting a Habitat for Humanity house:

During the selection of beneficiaries for new Habitat houses, we learned about Arputha Mary, a widow with an eight-year-old daughter, living in a small hut in Azhahappapuram Village. She earned her livelihood by doing agricultural labor and by weaving coconut leaves. The village committee members reported that she is a poor lady very much in need of a house. They said she would be prompt in making repayments. Arputha Mary belonged to the most backward class community. Even so, her family led a peaceful life.

Unfortunately, one day her husband, Mr. Paul Raj met with an accident and he succumbed to the injuries. After the loss of her husband, the young widow and her daughter led a very miserable life in a small thatched hut. A Habitat house was allotted to her. Then many well-wishers came to her aid. She and her friends succeeded in taking all the materials to the house site in two days. Bricks, sand, wood, and cement were all ready. The work was started on July 8, 1997, and the house was dedicated on September 15, 1997. So in seventy days the house was completed.

She, a poor widow, worked day and night to contribute her sweat equity. On the day of the dedication of the house she invited all her relations and neighbors and celebrated a small feast. All of those present were very much pleased to see such a house for this poor widow. She gave her first house payment on that date and from that date she has been regular in attending homeowners meetings and in paying the required house payments. She is now leading a peaceful and happy life in the Habitat house with her daughter and is very thankful to Habitat for Humanity.

Happiness Heart Attack

Another Habitat house brought incredible joy and great happiness to a family from the Philippines. The house was not built in the Philippines but rather in Cookeville, Tennessee, by Putnam County Habitat for Humanity.

Maria Gaerlan moved from the Philippines to the United States in 1987. In the years since, she has had to struggle to support and raise six children. She moved to Putnam County because her mother and stepfather lived there. But Maria lived in public housing and worked a variety of jobs. She long had the American dream of owning her own home.

Maria became a licensed medical lab technician in 1992. That work gave her better income but not enough to get conventional financing from a bank to purchase a house. She learned about Habitat for Humanity and applied for a house in 1997. In June of that year she was notified that her family had been selected as the fifth partner family for Putnam County Habitat. But Maria was told that her house would probably not be started until late 1998. Fortunately, J&S Construction Company in Cookeville decided to do a Habitat blitz-build as a way to celebrate forty years in business. A day was selected to do the blitz—December 12, 1997—and the house would be for Maria and her children.

This exciting information was kept secret from Maria. It was decided to do a surprise announcement at the South Central Regional Conference in Knoxville in mid-November. Maria attended that conference along with two of her daughters and Putnam County Habitat executive director Candace Thompson and vice chair Jane D. Reagor. Linda and I also attended that conference. We all sat at the same table that Friday evening. Jane and Candace had told us about the plan to build Maria's house, but they wanted to surprise Maria with a dramatic announcement from the podium. Jim Crowley, regional director for the South Central Region, made the announcement later that evening.

When Maria heard the news, she jumped up and exclaimed, "I'm going to have a heart attack!" Others jumped up with her and knocked the table over. Jane Reagor said, "These are the real moments in life when you experience and share true joy and happiness with your neighbors."

On December 12, J&S Construction, Smith Electric, and CHC Mechanical Services gathered at the site at 6 A.M. Less than ten hours later, Maria's house was finished, with only a few final touches to be done at a later date. Then, on Christmas Eve, at 8 A.M. with a Christmas tree in place, the dedication service for Maria's new home took place.

Maria exclaimed, "I thank the Lord above who has given all these people the thought of doing this. And it was just for me. I am blessed. I thank everybody, J&S, Habitat, and everybody in Cookeville who had a part in this. It is such a prize. What a Christmas. I will not forget this, all my life. It is awesome."

Fort Worth Smile

The happiness of Earl El-Amin of Fort Worth, Texas, was shown by the smile on his face. Earl, a single dad raising four children, is the proud owner of a house built by the St. Paul companies. When asked if his Habitat house had made a difference in his life and in the lives of his children, he spoke enthusiastically about how much more comfortable the family is in a house that is large enough for everyone.

Another blessing is that the house is close to his father, who takes care of the children while Earl works as a youth counselor with the Dallas County Juvenile Detention Department. Earl said that his family enjoys a peace of mind that they had never before experienced.

"My children are more stable and more secure, knowing where they will be next month, next year, and in the years to come. Owning a Habitat house has freed me to work on other things. I'm trying to go back to college. All of this is hard to explain," Earl concluded. "I just wish everyone could see the smile on my face."

Shoveling Happiness

A homeowner in Libby, Montana, expressed his happiness in an unusual way. Patricia Tackes of Habitat for Humanity Kootnai Valley Partners shared the story.

"The winter of 1996–97 gave us more snow than ever in our history, with crushed roofs and mountains of snow. One day a board member met the Osman family (one of our homeowner families) in the grocery store, and they chatted mainly about the weather. The board member said that he was really tired of having to shovel his driveway and walk. Mr. Osman replied interestingly, 'I am truly grateful to have a driveway and walk to shovel!'"

Awesomely Happy

Michelle Ewing of Irvine, California, was awesomely happy to be in her new Habitat house for many reasons. She and her two daughters, Chanelle and Shawna, moved into their dream house in late 1998.

Michelle is a single mother whose husband deserted her after six years of marriage and two children. She had not worked outside the home during the marriage. After the split, Michelle got her cosmetology license and briefly did manicures and pedicures to support herself and the girls. But since her teenage years, Michelle suffered from scoliosis. While still at home with her parents, she was in a full-body cast four times. She had several surgeries, one resulting in a steel rod in her back. She had seven outpatient surgeries in 1997 and 1998 and was in nearly constant pain. Eventually, she had to quit doing manicures and pedicures because of her back pain.

Michelle and her girls moved in with Michelle's mother. They stayed there for three years. Then Michelle rented a little apartment in a bad neighborhood right by a freeway. It was all she could afford. There were fights, drugs, shootings, a suicide, and repeated vandalism. One time, Michelle was assaulted on the premises. Soon, though, she became a teacher's aide. She also went back to school and studied sign language to become a teacher for the deaf.

Then Michelle heard about Habitat for Humanity from a friend. She went to a volunteer orientation and started volunteering for special events with Habitat for Humanity of Orange County. She did this for a year before applying for a house. Michelle is deeply religious and prayed a great deal about getting a decent place to live. She believes it was a miracle that she and her girls were chosen to have a new Habitat house. Michelle says that she asked the Lord into her life in 1988 on a freeway. She had previously abused alcohol but has been alcohol-free since February 16, 1988.

Michelle worked diligently with a host of volunteers to build her house. Finally it was finished and ready for her and her girls. She said that sometimes she feels like she is visiting someone else or is in a hotel when she walks into the new house. Then she remembers that the house is really theirs.

The family is now at peace. The girls are excited about having their own rooms and have decorated them. They relish the fact that they can now be friends with their neighbors and are not afraid to go outside. Michelle is so thankful to have a place to call home—her own happy home.

Volunteer Happiness

The last two stories in this chapter are not about homeowners but about volunteers and how their work with this ministry has given them meaning and happiness. The first person did not want his name used, and in a way, that is very appropriate. This anonymous volunteer speaks the language of tens of thousands of others who have given so freely to this work and who have not only blessed others tremendously but have been blessed even more. They derive enormous satisfaction, joy, and great happiness from the whole process.

This is what he wrote:

The beginning was April 1997. For a long time I had been looking for something to occupy my time. I was going to my first weekend of many to come working for the San Antonio affiliate of Habitat for Humanity. Having been told to arrive at the site around 8 A.M., I was surprised upon arrival to see all the people already there. It was quite a sight. Many homes were completed but new construction was scattered among them. I was amazed at the overall sight of people and houses. I reported to the sign-in table and was greeted by one of the Habitat members who had given me the orientation a few weeks earlier. She directed me to the site next to the sign-in table.

There were three men working there. I introduced myself and asked if they needed any help. The house leader replied "yes" with an offering of a handshake. Throughout the day others arrived.

I had obtained previous experience working in the construction trade with my brother, so I was able to give advice to the less-skilled volunteers. I noticed that all of the volunteers showed a special excitement about this project; they were all willing to take on the next task and learn. At the end of the first workday, I was so exhausted that I did not even consider doing it again.

During the following week, I pondered my experience at the work site. My sore joints from the weekend seemed to disappear. I remembered the people I worked with and the sense of pride that we all had at the end of the first day and decided that I really enjoyed what I had done.

The following weekend, I was at the site again. This time I got there early. I had time to sit on a bench built in the community park and watch the sun come up and contemplate my own life. It seemed like I had never done this before and it gave me a wonderful sensation of peace, even though I was going through troubled times. It was quite something to see people arrive and the crews gather at the house sites that were being built. It kind of reminded me of those fast-moving depictions of a city coming to life that we sometimes see on television with things moving so fast they nearly become a blur. Before I knew it, we were hard at work on the second day of the twelve-day schedule to complete this house. It is hard to believe that these houses could be built in twelve days, but after working on the first house, I developed a better understanding. Almost every Saturday of the twelve to come, we had a good turnout of volunteers.

Three months later, the house was nearing completion. Prior to the families moving into it, a dedication was planned. Many of the workers from Habitat, as well as the volunteers gathered, and the house received a blessing. After my first experience of this blessing, I again reflected on my life and found a joy in my heart for being able to help.

It has been several months since the start of the first house (I like to think of it as *my* first house), and I am still engrossed.

I have devoted many Saturdays to these homes and had the pleasure of meeting many new people who I believe share the similar joy that I have. I now know the staff and admire them for what they are doing. But most of all, I think I have followed my calling. I believe that I have received as much out of building these homes as those who receive them.

The Happiest Habitat Volunteer

This final story is about a truly amazing volunteer. Jackie Goodman, introduced in chapter 8, is not like the anonymous volunteer above; she is not a building volunteer in the traditional sense. But she is a "builder," and hers is an enormous success story. Jackie is also probably the happiest volunteer in the ministry of Habitat for Humanity anywhere in the world!

In 1990, Jackie founded the Habitat for Humanity in Atlanta Birdhouse Artfest. She gets artists in the Atlanta area to design and paint birdhouses that are then sold at an auction (silent and live) each fall. In the first year of the event, fifteen thousand dollars was raised for Atlanta Habitat for Humanity. Each year since, the artfest has raised more and more money. In 1998 the artfest raised the incredible sum of one hundred and fifteen thousand dollars. In all, Jackie has raised for Atlanta Habitat in excess of three hundred thousand dollars.

Since 1996, the Georgia quilt project has been a part of the artfest, adding more than twenty thousand dollars to the funds raised for building Habitat houses in Atlanta. In 1998 Georgia quilters sold seventy-five pieces and all proceeds went to house building.

Jackie Goodman displays samples of birdhouses that have raised thousands of dollars for Atlanta Habitat for Humanity.

Jackie, who makes her living selling houses through her real-estate firm, says about her hugely successful project, "I just love selling houses for people and birds!" We met in 1988. She has been a dear friend ever since. She always has a huge smile on her face. She bubbles over with joy, and her smile radiates a huge happiness. Every time I speak in Atlanta, or anywhere near

Atlanta, Jackie is in the audience, usually on the front row or near the front row. And she always has someone with her—someone she wants me to meet.

Jackie is the epitome of a fantastic Habitat volunteer. Joyous. Radiant. Positive. Effective. Happy. She inspires me and thousands of others who know about her fantastic work. Jackie has inspired numerous other birdhouse sales in other cities that have produced thousands more dollars for Habitat houses. That makes me and a lot of other folks happy.

Happiness. This hammer-and-nail ministry brings so much to people, and it means so many different things to various individuals. For certain, though, the work brings loads of happiness to just about everyone involved in the process—homeowners, volunteers, donors—everyone.

CHAPTER 14

Building Habitat's Future

*Students are a lifeline to Habitat for Humanity. Many campus
chapters members go on to found or become board members of
Habitat affiliates. As high as 50 percent of the alumni,
I'm certain, stay active on a local level.*
—Sonja Lewis, former Habitat campus chapters department director

Habitat for Humanity will be twenty-four years old in the year 2000. For several years I have believed that a new social and religious movement is forming to rid the world of substandard houses and homelessness. Habitat for Humanity is in the vanguard of launching and nurturing that movement.

In our first twenty-four years, as stated above, we will have built a hundred thousand houses. In 1998, we dedicated the seventy-thousandth Habitat house. In 1999, the milestone eighty-five-thousandth house was built. In September 2000, during our annual Building on Faith week, the one-hundred-thousandth house will be built and dedicated. We will have housed at least half a million people. According to our projections, the next one-hundred-thousand houses will be built in just five years. That means that we will be ready to dedicate the two-hundred-thousandth house in 2005, and those houses will be providing shelter for a *million* people.

At the time of this writing in 1999, Habitat for Humanity has affiliates in more than fifteen hundred towns and cities in all fifty of the United States, and we continue to add affiliates at the rate of five to ten each month. Around the world we now operate in more than sixty nations, and more countries are joining the Habitat ranks every year.

Our plan is to push steadily to install Habitat affiliates in more and more countries until we have a presence in every nation on earth. Then within each country we steadily push to take the program to every section of the country, always building houses, of course, but planting the powerful idea that everyone should have at the very least a simple, decent place to live.

The momentum in this work is strong and getting stronger all the time. Even so, the task before us is enormous. The United Nations estimates that more than a hundred million people in the world are homeless and that more than one and a half billion people live in poverty housing.

Even with a growing movement in place, it is going to take a very long time to accomplish the goal of ending poverty housing and homelessness. That's why it is essential to engage the next generation and the one after that and the one after that in this ministry of hammering out faith and love.

And that is precisely what is happening! Tens of thousands of young people are now actively involved in Habitat for Humanity, and they are building hundreds of houses and a whole lot more.

"Habitat for Humanity is the one volunteer organization where you can see right then that what you do makes a difference," said Tarra Wright, a Texas Tech University student and member of a campus chapter.

"I enjoy working with the young men and women. They have a great enthusiasm and energy," said Howard Yandell, First United Methodist Church coordinator for the building project that Tarra and her chapter member friends helped build.

These sorts of responses can be heard across the country, from campus to campus, and from everyone involved with these student Habitat enthusiasts. This wonderful involvement of young people in Habitat for Humanity started with a phone call.

A Phone Call

In 1987, Gary Cook called to invite me to speak at Baylor University. Now president of Dallas Baptist University, at the time Gary was the director of denominational and community relations and the assistant to the president at Baylor. He had known about Habitat for Humanity for some time, and he was personally involved in the new Habitat affiliate in Waco, Texas.

But Gary had more than a speaking engagement in mind when he called. He asked me if we had any Habitat organizations at colleges or universities. When I replied that we did not, he responded by saying that he would like to take the lead in organizing the first student group at Baylor.

I thought the idea was terrific. If we could form student chapters of Habitat for Humanity all over the country and eventually in other countries to raise money and awareness and do hands-on construction, the impact could be incredible on the entire movement's future.

I went to Baylor and spoke at a chapel service that was attended by more than two thousand students. When I presented the idea of Baylor's forming the first campus chapter of Habitat for Humanity *in the world,* the students responded very positively.

There was an official commissioning of the chapter. An outstanding student, Ward Hayworth, was named first chapter president. Baylor president Herbert Reynolds and several faculty and staff became charter members, along with two hundred students.

The new chapter plunged immediately into work with Waco Habitat for Humanity. The students raised funds, spread the word about the ministry on campus, helped build Habitat houses locally, and also organized work groups to go to other affiliates in the United States and Central America.

Back in Americus I set up a campus chapters department and named David M. Eastis, a dedicated and energetic young man from San Jose, California, to be the first director. David very quickly got the new department organized and started spreading the word about the new program to colleges and universities across the country. I talked about the new thrust of the work in practically all of my speeches, and I also wrote hundreds of letters promoting this exciting new idea.

I wanted to see chapters form, at the very least, in the same cities as our local affiliates in order to assist them. Each chapter would have a student president, but anyone on campus could join, including staff and faculty. These staff and faculty people could also hold one or more offices, such as secretary or treasurer, to ensure continuity over the years.

The money raised by the campus chapters should be turned over to the local affiliate or sent to Habitat for Humanity International. The campus chapters would not build or renovate houses except in partnership and with the full concurrence of a local Habitat affiliate.

The chapters would provide volunteer workers to the affiliate in the immediate vicinity of the college or university to build, renovate, or repair houses for needy families or help the affiliate in other ways. The chapters could also organize work teams to travel to distant sites in the United States or in some other country to help a building project. The chapters would be encouraged to recruit students to work full time with Habitat following graduation or to continue their involvement with the ministry as part-time volunteers.

Students would also be encouraged to start new Habitat affiliates if, after graduation, they moved to a town or city that had not yet launched an affiliate. That very thing happened quite soon. Two years after starting the campus chapter at Baylor, I was in Little Rock, Arkansas. They were just starting a new affiliate. When I asked how the affiliate got started, I was told that a couple of former Baylor students had taken jobs in town after graduation and they had planted the idea for the affiliate.

Therein lies our future.

Household Word

One of the primary goals of every campus chapter is to make Habitat for Humanity a household word. The official motto of campus chapters is, "Love should not be just words and talk; it must be true love, which shows itself in action" (1 John 3:18 TEV).

Only a year after that first chapter was founded at Baylor, a goal was set to have at least ten chapters formed and operating by the time of the twelfth anniversary celebration of Habitat for Humanity International, which was set for mid-September 1988 in Atlanta. Amazingly, at the celebration we were able to announce not ten chapters, but thirty-six! Jimmy Carter, keynote speaker for the celebration, presented charters to representatives of the thirty-six colleges and universities. The ball was definitely rolling.

The very next month we accepted the first high school chapter—Marist School, a Roman Catholic day school in Atlanta for

Lauren Perry from the University of Dayton participates in Building on Faith week in Canton, Ohio, in 1998. Photo by Kim MacDonald.

young men and women in grades seven through twelve. Other high school chapters quickly followed.

That same month the first chapter was formed at a seminary. Actually, the chapter was at a college and seminary—Bethel College and Seminary in St. Paul, Minnesota. Four months later, the second seminary chapter was formed at Asbury Theological Seminary in Wilmore, Kentucky.

In 1989, David McDaniels became the second director of the campus chapters department. David came to us from Ithaca College in New York where he had been the Protestant chaplain and organizer of the school's campus chapter. When David assumed leadership of the department, there were seventy chapters in twenty-five states and the District of Columbia. A year later, there were a hundred chapters. When David left the department in 1993, there were two hundred and seventy-five chapters.

David's greatest contribution to the campus chapters program was his idea to have a Collegiate Challenge Spring Break Building Blitz. He recruited students from twenty-six chapters to build houses in Coahoma, Mississippi, during their spring break in 1990. These students raised sixty thousand dollars, enough to build six houses in Coahoma, and another six thousand dollars to build six more houses overseas. One house was built in Coahoma every week for six weeks. New groups of students arrived each week to build another house. Dedications were held each Friday.

Every year since 1990 the program has grown. In 1994, more than forty-three hundred students took part in the collegiate challenge spring break blitz-build. In 1995, the number of participating students topped five thousand.

In 1999, more than seventy-five hundred students went out from more than five hundred schools to work with one hundred and fifty affiliates all over the United States. The students raised in excess of seven hundred thousand dollars for their work.

Sonja Lewis became the third director of the campus chapters department. Under her leadership, from 1993 to 1997, the program grew to 478 chapters in forty-six states and eight foreign countries. Sonja also expanded the department to include younger children. In February 1996 the department was renamed the Campus Chapters and Youth Programs department.

Sonja organized and directed the observance of the tenth anniversary of campus chapters work. That very significant event was held at Baylor University on November 21 to 23, 1997. Students from twenty-four colleges and universities gathered for the celebration. The theme was Building

Tomorrow—Starting Today. Linda and I were privileged to attend. Gary Cook was also with us for that historic celebration.

A highlight of the event was the launching of a one-hundred-house build by campus chapters. The first of the hundred houses was already being built by the Baylor chapter in partnership with Waco Habitat for Humanity. A dedication service for that first house was held as the concluding event on Saturday afternoon. Over the next twelve months, one hundred houses were to be built by campus chapters across the country and in other nations.

A Hundred Houses—Five Hundred Chapters

And those houses were built. I was present to dedicate the one-hundredth house in DePere, Wisconsin, in November 1998, built by the campus chapter of St. Norbert's College in partnership with the local Habitat affiliate. That chapter, incidentally, was designated as the milestone five-hundredth campus chapter. The designation, interestingly enough, was made in the summer. By the time I got there in November, more chapters had been approved so actually, as of that date, there were five hundred and twenty-three campus chapters in all fifty states and nine other countries. But the St. Norbert's campus chapter was *exactly* the five-hundredth one in the United States. More than one hundred and seventy students worked on the house, and they were so proud of what they had accomplished. They were beaming at the dedication service.

Students Are a Lifeline

"Students are a lifeline to Habitat for Humanity," Sonja Lewis told a group recently. "Many campus chapters program members go on to found or become board members of Habitat affiliates. As high as 50 percent of the alumni, I'm certain, stay active on a local level."

Former campus chapter members also remain involved in other ways. Most of the staff members and volunteers in the CCYP department in Americus are young people who were involved in a campus chapter before moving to Americus to join the department. Two examples are Kevin Gilmore and Erica Karlovits.

Kevin is from Milford, Delaware, and attended Elon College in North Carolina. Quite active in the campus chapter, he helped the local affiliate find land for building more houses, traveled to Homestead, Florida, to help rebuild houses following Hurricane Andrew and led work crews several times during his college career to various sites.

After graduation, Kevin served for a year as a campus chapters outreach coordinator. He helped one hundred and seventy campus chapters in thirteen states develop and strengthen their programs. Next, Kevin trained to be a Global Village team leader. He then led a group of Elderhostel volunteers on a short-term mission trip to Guatemala.

Following the Global Village excursion, Kevin trained in Americus and Costa Rica to become an international partner. He arrived in Guatemala in September 1998 and is now serving in the national office in Quetzaltenango. He is training local Habitat people to be better leaders in their respective areas around the country. There are now one hundred and thirty building sites in that country.

Erica Karlovits was very active in the campus chapter at the University of Connecticut. During her senior year, she served as president of the chapter. After graduation, she came to Americus. For the first year, she worked in the campus chapters and youth program department as an outreach coordinator for the West Coast. In August 1998, she joined the Global Village department as a work camp coordinator. In that capacity, she not only coordinated work camps directed by other people, she led work teams herself. Both in the summer of 1997 and 1998, she led groups to Zambia. Her example and dedication were well known back at the University of Connecticut. The chapter there raised money for her good work. Two Raise the Roof concerts during the 1997 and 1998 school terms raised six hundred and fifty dollars for the ongoing work in Zambia.

International Campus Chapters

In January 1998, Jerome Williams became the fourth director of the department. He took the campus chapters program to new heights in the months ahead including breaking a new record on participation in the collegiate challenge spring break blitz-build.

The first campus chapter outside the United States was at the University of Technology in Lae, Papua, New Guinea. It was organized in 1990. In the years since, other chapters have formed in Zambia, Poland, the Philippines, and several other countries.

Outside the United States, the work has taken deepest root among Canadian schools. There are currently ten campus chapters in that country.

One of the most active campus chapters in Canada is at Wilfrid Laurier University in Waterloo, Ontario. That dynamic chapter not only works with the local affiliate in Waterloo but also tithes to Habitat for Humanity Canada to support a sister Habitat program in Jamaica. The chapter also sends work teams to Jamaica. In February 1999 nineteen students traveled to Jamaica to work at the new affiliate in Montego Bay.

Wilfrid Laurier campus chapter members have also organized work teams to work in Cleveland, Ohio; Grand Rapids, Michigan; and New Orleans, Louisiana.

Joanna Kleuskens, national coordinator of campus chapters in Canada, said, "Students play a vital role in the organization. Campus chapters are a big boost to the work of Habitat. Students provide much-needed ideas, energy, and manpower to local affiliates." The campus chapter at the University of Alberta in western Canada set a goal that in 1999 one of the houses built by the Edmonton affiliate would be sponsored by the chapter.

Such Incredible Energy

Wilmer Martin, president of Habitat for Humanity Canada, says that their goal is to have fifteen campus chapters in Canada by the beginning of the year 2000. Wilmer is most effusive in his praise and excitement about the growing work among students in Canada. He exclaimed, "Young people bring such incredible energy to our affiliate programs. When young people get filled with enthusiasm for Habitat, it goes with them for a lifetime."

There are more than five hundred and fifty campus chapters today around the world and more forming every month. In the United States, there are 3,706 two- and four-year colleges and universities. We intend to form campus chapters in all of them. In addition to these institutions of higher learning, there are thousands more junior and senior high schools. We want to organize campus chapters in as many of them as possible. We also want a presence in the lower grade schools.

Kids for Habitat

Even very young children can be involved and supportive of Habitat. For example, twelve-year-old Cynthia Landon in Hammond, Louisiana, formed Kids for Habitat. Nine children, ages eight to fourteen, raised more than a thousand dollars from a garage sale, a bake sale, and the sale of an old computer to sponsor a Habitat house in Honduras. A second garage sale produced $1,617 for Habitat houses. When Cynthia's family moved to Kansas, she started a Kids for Habitat there. In South Carolina, Sea Island Habitat for Humanity at John's Island started a Girl Scout troop for local Habitat homeowners' daughters. The troop, with more than twenty girls aged thirteen to eighteen, held numerous fund-raisers to support the local work of Habitat and to earn a recognition patch for community service.

AmeriCorps

Young people of all ages are otherwise engaged in the work of Habitat for Humanity in ways other than through a campus chapter, Kids for Humanity, or a Girl Scout or Boy Scout troop. Many work directly with local Habitat affiliates or as a part of a church work group.

Hundreds of hardworking young people have worked with Habitat affiliates all across the United States over the past several years as part of the AmeriCorps program. That innovative initiative, set in motion by President Clinton, has been a tremendous blessing to scores of Habitat affiliates. And the impact of those AmeriCorps workers grows stronger year after year.

Beyond North America

We want to vigorously launch campus chapters, certainly in every country where there are Habitat affiliates but even in other countries. Four campus chapters were started in Japan in 1997, a country where there are no affiliates. The students, though, are incredibly excited about the ministry. They raise money and recruit volunteers to work around the world, primarily the Philippines.

Campus chapters are making a powerful difference in hundreds of ways in the ever-escalating venture of Habitat for Humanity. The chapter at Mercer University in Macon, Georgia, for example, builds a house every year with the local affiliate. The chapter at Elon College, North Carolina, does the same.

The Lynchburg College chapter in Lynchburg, Virginia, completed its fifth Habitat house in 1998 in connection with the chapter's tenth anniversary celebration. The house had special significance for several reasons. It was the first house that the students independently sponsored. It was also to be the house for Gertrude Brown, a dining-room employee at the college for more than twenty years. And the house was an Oprah Angel House.

Members of the Erskine College campus chapter in Due West, South Carolina, became good friends with Gladys Holmes while helping to build her house in Greenwood, South Carolina. During spring break in 1997, Gladys took a week off from her three jobs and joined sixteen students on their collegiate challenge trip to Montgomery, Alabama.

Habitat Bicycle Challenge

One of the most exciting activities of a campus chapter, and certainly the most ambitious and lucrative, is the annual Habitat Bicycle Challenge. Started by students of the Yale University campus chapter in New Haven, Connecticut, to raise money for the New Haven affiliate, the event has grown steadily every year. Students give up eight weeks of their summer to bike across the country to raise money and awareness for Habitat.

Yale University student Antony Brydon came up with the idea after spending a semester at Habitat headquarters in Americus. In 1994, Antony and seven other riders biked from New Haven to San Francisco in three months. They covered more than five thousand miles on that first long journey.

The new route after that first summer became Washington, D.C., to San Francisco. The challenge became so popular that it was expanded to two routes in 1998—Washington to San Francisco and New Haven to Vancouver, British Columbia. That year, sixty-four riders joined the effort, representing thirteen different colleges and universities. Those sixty-four riders raised more than one hundred thousand dollars for the work of Habitat for Humanity.

All cyclists are required to raise a minimum of twenty-five hundred dollars each. Since there are corporate sponsors to help cover expenses and because housing and food are freely given by churches and other organizations along the way, all money raised by the cyclists goes directly into building Habitat houses.

Homecoming Float

Campus chapter members at Wayne State College in Wayne, Nebraska, dressed up as homeless people around a fire barrel on a homecoming float and passed out hundreds of flyers about Habitat for Humanity and their chapter. The chapter regularly helps the nearby Norfolk affiliate. Aaron Truex, vice president of the chapter, said that he'd never forget the day they worked on one house's landscaping and stucco work. He explained that they also moved a lot of dirt in preparation for the sidewalk, driveway, and patio. Aaron said that the work was "not glorifying in any way, but that it was gratifying."

International Campus Partners

A very special partnership has been developed between the campus chapter at the University of North Carolina–Chapel Hill and campus chapters at two secondary schools in Siquatepeque, Honduras—Pablo Weir and Genaro Munoz Hernandez Schools. The relationship began in 1996 when the campus chapter from North Carolina asked to work with local chapters when they went on a Global Village trip that year. So, arrangements were made for students from Pablo Weir and Genaro Munoz to join them on the work site in Siquatepeque.

University of North Carolina-Chapel Hill students joined with the Siquatepeque Habitat for Humanity affiliate in building houses in Honduras.

In 1997, the second year of the partnership, the students worked on three houses. In the evenings they shared meals with people in the community. In addition to their work with the local affiliate, the students visited the ancient Mayan ruins in Copan and went sightseeing at other interesting places.

The UNC chapter raised seventeen thousand dollars for the expenses of the fifteen participants and the student-sponsored house and one thousand dollars as starter funds for next year's team. Seven thousand dollars was raised by letters requesting contributions from friends and family. Donors were asked to buy shares with a promise of a postcard from Honduras. The chapter also held a Honduran dinner, sold spaghetti dinners, provided baby-sitting services on Friday nights, stuffed envelopes for a local company, and operated a concession stand at basketball games.

The 1999 trip to Honduras was the fourth for the UNC chapter. That excursion took on added meaning and urgency because of the devastating hurricane that struck the country in late 1998.

Another natural disaster, a tornado, greatly affected a campus chapter and a college earlier in the same year. On March 28, one hundred and twenty-five students from Gustavus Adolphus College in St. Peter, Minnesota, set out in six teams for their collegiate challenge trips. Most arrived at their destinations the following evening, Saturday, March 29, in Sisters, Oregon; Hartford, Connecticut; Oklahoma City, Oklahoma; Norristown, Pennsylvania; Christianburg, Virginia; and Columbus, Georgia.

As they were arriving at their destinations, the students learned that a devastating tornado had hit their college campus and town. St. Peter's was in shambles. Windows in the dormitories were broken and rain poured in. Most trees on campus were destroyed. One student's car was crushed by a tree.

Should the students return to the college to help out with the incredible destruction? Many prayers went up. Prayer circles were formed, and students prayed aloud asking for guidance and giving thanks also because most students were away on spring break. After heated discussion and with a guarantee from school officials that plenty of work would remain at the end of the week, all one hundred and twenty-five students decided to stay for their week of work with the various local affiliates. All those who worked with the students that week were impressed and proud of their commitment—as was I.

Bringing Foreign Partners Together

Campus chapters are even breaking barriers between countries with long histories of conflict by bringing the countries' youth together toward a common goal. That was the case when campus chapters from several prominent Asian countries came together to help the Philippines.

"Before we began the work camp, we were Korean, Japanese, and Filipinos. By the end of the work camp, we were one big family in Christ," commented Professor Akiie Ninomaya of Kwansei Gakuin University in Japan.

All of these students were at a short-term mission in Tarlac, the Philippines. More than twenty-seven students from three countries, speaking three different languages, learned a lot more than just how to build a house.

Volunteers take a break from house building in the rain near Tae'baek City, Korea.

Anita Mellott, editor of the *International Affiliate Update* for Habitat, wrote about that experience. "Perhaps the best indication of the barriers being broken through Habitat was during 'Friendship Night,' an evening of fun and fellowship toward the end of the work project," she wrote. "A Japanese student shared an incident that took place during the build. While filling bags with volcanic ash, a rock flew out of a Korean's hand and hit a Japanese student on the shin. 'I did not feel the pain,' said the Japanese student, 'because of the love and concern that my Korean brother showed me. A wound is nothing. All I saw was how he cared for me.'"

Pioneering Partnership

A pioneering and very creative partnership was forged in the small town of Springfield, Georgia, in late 1997. The students in a construction class at

Effingham County High School built a Habitat house on the school grounds for Effingham Habitat for Humanity. Materials for the house were donated by the Georgia-Pacific Foundation and its Savannah hardwood-plywood plant. Kenny Lay, ECHS construction teacher and a member of the Habitat affiliate board, put the deal together and managed the construction of the house. Some forty students worked on the house. When the eleven-hundred-square-foot house was completed in 1998, it was hauled to a nearby lot and finishing work done so that the new homeowner family could move in a few months later.

Doug Peavy, a fellow Habitat volunteer and board member, manager of the Georgia-Pacific Savannah Plant and also very involved with Kenny Lay in this venture, said, "The idea is to fulfill someone's dream for owning a home, while giving the students invaluable training in building a major structure and a sense of responsibility for their fellow men."

The experience of everyone involved in the undertaking was so positive that a second house was started in September 1998. Again funding was provided by Georgia-Pacific. The house will be completed in 1999.

Because of the great success with the first house, a campus chapter was formed at the nearby high school. I met several of the members of the chapter at the Southeast Regional Conference in Jacksonville, Florida, in October 1998 when I was there with Linda to speak at the Habitation service and to lead a couple of workshops. The conference brought together Habitat people from Georgia, Florida, and Puerto Rico. I posed for a picture with the students. All the while they were excitedly telling me about building that second house.

Love into Action

As the above stories reveal—and these are but a tiny fraction of the *whole* story of campus chapters—students are contributing so very much to the ministry of Habitat for Humanity. Early in the development of the campus chapters program, Duke University campus chapter president Sally Higgins spoke eloquently about the meaning and significance of the work. "Campus chapters of Habitat for Humanity provide students with the opportunities to put their love into action," she said. "And students are increasingly proving that they are idealistic enough to dream, compassionate enough to care, and responsible enough to act."

Students by the tens of thousands are now regularly building and renovating Habitat houses, all across the United States and around the world.

But they are doing so much more. Again as the above stories reveal, the students are raising money, spreading the word, finding land on which to build, adding loads and loads of enthusiasm and energy, and so much more.

Sociology, Geography, Anthropology, Spirituality, and Construction

The students are also gaining in the whole process. As one student leader from the University of New Hampshire campus chapter explained, "Expectations were to build a house and work with a family. Much, much more happened because we all learned sociology, geography, anthropology, religion, spirituality, and construction."

The most significant impact of the campus chapters program is on the *future* of this faith-and-house-building venture. These young people are ensuring a future for this movement. They are the future. They are building houses by the thousands. They are bringing action by the truckload. They are lavishing love and enthusiasm to other volunteers, Habitat staff, people in the neighborhoods where they work, and to the new Habitat homeowners.

These students are dedicated, too, as the episode with the Gustavus Adolphus students revealed. They are also bringing new ideas to the table. In short, they are an incredible blessing. They are loved and appreciated—and so much more.

Our Future

All of the chapters in this book show how so much more than houses is being built in this ministry by everyone involved right at this moment. But I chose to conclude *More Than Houses* with a chapter on Habitat's youth partners for all the reasons set forth above. Young people are guaranteeing that the work will not only continue but will also accelerate in the years ahead.

Sally Higgins said it best: "Students are idealistic enough to dream, compassionate enough to care, and responsible enough to act." Dreaming, caring, acting. That means houses and more houses. Love and more love. Faith and more faith.

More than houses?

Yes!

Nathaniel and Emily Potter, children of Habitat
volunteers in Mason, Michigan, demonstrate
that Habitat's future is bright.
Photo by Kim MacDonald.

United Nations
Habitat Agenda

There are three basic themes in the United Nations Habitat Agenda: the right to housing ("adequate shelter for everyone"); sustainable development; and the "enabling" approach.

The Universal Declaration of Human Rights named housing as an integral part of the right to an adequate standard of living in 1948. However, the right to housing was not a legally binding treaty obligation until 1969, when the International Convention on the Elimination of All Forms of Racial Discrimination forbade racial discrimination in the realization of the right to housing. Since that time, several international treaties have created the right to housing as a binding obligation in international law. Of these treaties, the United States has ratified only one—the International Convention on the Elimination of All Forms of Racial Discrimination. The treaty, though, is not part of U.S. law since no legislation has been enacted to implement it.

The Habitat Agenda (which uses "shelter" interchangeably with "housing") incorporates *the right to housing* as a basic principle, spelling out the elements as follows: "Adequate shelter means more than a roof over one's head. It also means adequate privacy; adequate space; physical accessibility; adequate security; security of tenure; structural stability and durability; adequate lighting, heating and ventilation; adequate basic infrastructure, such as water supply, sanitation, and waste-management facilities; suitable environmental quality and health-related factors; and adequate and accessible location with regard to work and basic facilities; all of which should be available at an affordable cost" (The Habitat Agenda, Chapter IV, para. 60).

The *sustainable development* section of the agenda is about the broader problems of urban and nonurban areas. It includes housing but puts it in a larger context of social and economic development, including a concern for the environment.

The enabling approach states that it is up to the government to ensure that housing policy is implemented within the human rights framework and has special responsibility to groups of people who are excluded from the housing market. But the agenda also emphasizes that securing housing for everyone is not solely dependent on the state. Governments are called upon to enable and promote public-private partnerships for housing development and services.

Address to the Plenary

A Simple, Decent Place to Live
UN Habitat II Conference
Istanbul, Turkey
June 5, 1996

MILLARD FULLER

T hank you, Mr. Chairman. It is an honor to address all of you gathered here today on the occasion of this historic Habitat II Conference. I am deeply grateful to Dr. Wally N'Dow and other Habitat leaders who have labored diligently for many, many months to plan this conclave and especially grateful that they have so completely included Habitat for Humanity and other NGOs in this event.

The task at hand—namely to assure adequate shelter and livable, sustainable communities that nurture and enhance life rather than demeaning and destroying it—is too big, too daunting to leave any potential ally standing idly on the sidelines.

Every such potential ally from whatever realm, governmental or otherwise, should be encouraged to make the maximum contribution possible to help alleviate the suffering of our fellow human beings who are languishing in miserable living conditions. We can ill afford the luxury of leaving any of them on the sidelines of our noble struggle to provide adequate shelter for all.

Habitat for Humanity is certainly not on the sidelines. We are engaged. At work. Driving nails. Putting up walls. Capping off roofs. Developing communities. Building people—in two thousand locations in more than forty nations around the world.

And we are together. Habitat for Humanity feels a kinship with United

Nations Habitat. Actually, we started out together. Our common journey began in 1976. I was returning to the United States with my family from three years of building houses for needy families in Zaire, Africa. I remember listening to Voice of America broadcasts of the Habitat I conference in Vancouver, Canada. The word *habitat* was not used much in those days. It was not in vogue. But I was intrigued by the word and by its inclusive and full-bodied meaning in regard to human settlements.

My wife, Linda, and I returned to rural southwest Georgia to a small Christian community where we had previously built some houses for needy families. We had established a fund there called the "Fund for Humanity." Donations and noninterest loans were provided by individuals, companies, churches, and other nongovernmental organizations.

We gathered a group of people together to discuss the idea of forming an organization to build houses for the poor throughout the United States and in developing countries. The mission would be to put the tragedy of poverty housing and homelessness on the hearts and minds of people everywhere. The goal would be to eliminate poverty housing worldwide. We considered what to call the new organization. I remembered listening to the Habitat I conference. It was decided to put the two terms "Habitat" and "Fund for Humanity" together. Hence, the name of this new organization: Habitat for Humanity.

Today this organization that began so small twenty years ago is very much alive worldwide. It is providing solutions to problems of inadequate housing and homelessness. We are part of the United Nations Habitat family. We are, in a real sense, an offspring of Habitat I. And our very destiny is linked with this Habitat II Conference. Like members of a family, we look to each other for hope, for nourishment, and for new approaches. We recognize that the cries and anguish of more than a billion people who live in inadequate housing or no housing at all must be addressed with action. Today. At this moment in time.

At this gathering we must come up with more solutions and gain more allies in our resolve to ensure adequate housing for all people throughout the earth. We never know who will be listening. It could be someone who will help make dramatic progress in the twenty-first century. It might be someone who will be inspired by our deliberations and actions to undertake a magnificent new housing initiative in Africa, South America, or some other area of the world.

We can be grateful for what has been accomplished, but we must do more than double past efforts to successfully meet the awesome challenge at hand.

Habitat for Humanity began the size of a mustard seed. But by the end of 1996, we will have affiliates in more than thirteen hundred cities in the United States and in nearly nine hundred locations in forty-seven other countries. We will complete our fifty thousandth house in September. Our pace of building currently is more than twelve thousand houses per year. We have provided decent, affordable housing for more than a quarter of a million people. But it's not enough. We are an impatient offspring. We know that God is impatient with a global situation that has millions of people in urban and rural areas living in poverty and squalor that almost defies description.

Our contribution to this worldwide tragedy is very straightforward. In Habitat for Humanity, we follow basic principles. We depend on volunteers. Volunteers who care about their fellow human beings in need. Volunteers who give their money, their time, and their God-given talents. These volunteers form partnerships with each other, with churches, with other organizations, and with people in need. This partnership disregards social status, gender, skin color, cultural differences, faith differences. We call this "the theology of the hammer." The theology of the hammer brings together all kinds of people, all kinds of organizations, all kinds of churches and religious traditions, and all kinds of institutions to build and renovate houses with and for needy families. We are a Christian organization, but not exclusive and not doctrinal. Our desire is to put faith and love to work. To make a difference. To change things and change people for the better.

We build and renovate houses with volunteer labor and, wherever possible, donated materials. We require that homeowners work on their own houses. We call this sweat equity. We require homeowners to pay for their houses. We call this self-empowerment and self-help. We keep the houses affordable by not making any profit and not charging interest. The payments we receive are placed in a Fund for Humanity and recycled to pay for building more houses.

We ask that decisions on construction and family selection be made at the local level. Wherever possible, we use local materials and local construction expertise. This helps to keep the costs down. We insist that the houses be of simple design, decent and durable. We don't accept government funds for building houses. We do accept government funds to help acquire land or prepare infrastructure. Each project begins with a thankful prayer to God. Each day begins with a devotional message. We complete each house with a dedication and thanks to God. We know the true source of our vision and our ministry. There is nothing complex about this approach. It is open to everyone everywhere.

Habitat for Humanity also is open to new ideas in construction. We have a Department of the Environment that is taking us to new areas such as construction of straw-bale houses, fly-ash wall panels, and steel-frame structures. We are committed to showing respect for our environment and natural resources.

We minister to the poor, but also to the rich. We give them hope. We offer the rich a way to share their wealth. To be partners with the poor and to see the tangible, visible results of their generosity and compassion. Habitat for Humanity has built a bridge. The wealthy people of this world can cross this bridge to reach the poor and become partners with them. And the poor can cross the bridge, too, giving back with their house payments, modest donations, ideas, and volunteer work.

Young people are dynamically involved in Habitat for Humanity. We currently have campus chapters at four hundred colleges and universities in the United States and several other countries. Tens of thousands of students are raising millions of dollars and building hundreds of Habitat for Humanity houses. We also are organizing prisoners to build Habitat houses. And senior citizens are fully engaged. Indeed, our volunteers know no boundaries of age, race, or social standing as we build around the world.

We have come to Istanbul to share our journey with Habitat II. We believe God is blessing our humble efforts. Some will not embrace our approach. But all of us can and do, I believe, agree on this: Regardless of religion, tradition, or differing political persuasions, everyone should have a decent place to live on terms they can afford to pay.

In my religious tradition, we talk about grace. Grace is a gift from God. The opposite of grace is disgrace. Disgrace is what we see when we see homeless people or families living in miserable shacks or cardboard boxes or under trees and bridges or under no shelter whatsoever! A few years ago I met an eight-year-old girl named Jamie. She and her family had been living under a bridge. Today they have a Habitat for Humanity house. In my country, with all its wealth, we have many other people like Jamie—in fact, living in even worse conditions. It's a disgrace. It is unacceptable. It is unconscionable. All little girls and boys and their parents and other relatives need a decent place to live. We've got work to do.

I want to issue a personal call to eliminate poverty housing. I feel privileged today to add my voice and my appeal and, yes, my impatience to that of others. Let us stand united to declare that miserable living conditions are unacceptable. Whatever a person's gender, political party, religious commitment, or lack of religious commitment may be, that person's tenure on earth should be characterized by love, grace, compassion, and a decent place to live. That changed condition is not going to happen by accident. It will only happen when we put love into our work and join together. That coming together builds unity and makes us all strong for the continuing struggle.

In closing, I want to share with you my favorite poem, "Be Strong," penned by a Presbyterian pastor in New York City at the turn of the last century. His words are equally applicable today.

Be Strong!

We are not here to play, to dream, to drift;
We have hard work to do, and loads to lift;
Shun not the struggle-face it; 'tis God's gift.

Be Strong!

Say not, "The days are evil. Who's to blame?"
And fold the hands and acquiesce—oh shame!
Stand up, speak out, and bravely, in God's name.

Be Strong!

It matters not how deep entrenched the wrong;
How hard the battle goes, the day how long;
Faint not—fight on! Tomorrow comes the song.

Thank you my distinguished colleagues, sisters, and brothers of Habitat II, fellow citizens of the world who believe strongly in the simple but profound idea that everyone on earth deserves, at least, a simple, decent place to live. If we labor on and not grow weary, we will gain strength and tomorrow will bring the song.

Attachment A
Chapter Covenant

Covenant Between Dallas Area
Habitat for Humanity, Inc. and the ___ Chapter

Preface

Habitat for Humanity International, Dallas Area Habitat for Humanity, Inc. ("*Dallas Habitat*") and its chapters work as partners in an ecumenical Christian housing ministry. Dallas Habitat and its chapters work with donors, volunteers, and homeowners to create decent, affordable housing for those in need and to make shelter a matter of conscience with people everywhere. Although Habitat International will assist with information resources, training, publications, prayer support, and in other ways, Dallas Habitat is primarily and directly responsible for the legal, organizational, fund-raising, family selection and nurture, financial, and construction aspects of the work.

Mission Statement

Habitat for Humanity works in partnership with God and people everywhere, from all walks of life, to develop communities with people in need by building and renovating houses so that there are decent houses in decent communities in which people can live and grow into all that God intended.

Foundational Principles

1. Habitat for Humanity seeks to demonstrate the love and teachings of Jesus Christ to all people. While Habitat is a Christian organization, it invites and welcomes affiliate board and committee members,

volunteers, and donors from other faiths actively committed to Habitat's mission, method of operation, and principles. The board will reflect to the extent possible the ethnic diversity of the area to be served.

2. Habitat for Humanity is a people-to-people partnership drawing families and communities in need together with volunteers and resources to build decent affordable housing for needy people. Habitat is committed to developing and uplifting families and communities, not just to the construction of houses.

3. Habitat for Humanity builds, renovates, and repairs simple, decent, and affordable housing with people who are living in inadequate housing and who are unable to secure adequate housing by conventional means.

4. Habitat for Humanity selects homeowner families according to criteria that do not discriminate on the basis of race, creed, or ethnic background. All homeowners contribute sweat equity; they work as partners with the affiliate and other volunteers to accomplish Habitat's mission both locally and worldwide.

5. Habitat for Humanity sells houses to selected families with no profit or interest added. House payments will be used for construction or renovation of additional affordable housing.

6. Habitat for Humanity is a global partnership. In recognition of a commitment to that global partnership, each affiliate is expected to contribute at least 10 percent of its cash contributions to Habitat's international work. Funds specifically designated by a donor *for local work only* may be excluded from the tithe.

7. Habitat for Humanity does not and will not accept government funds for the construction of houses. Habitat for Humanity welcomes partnership with governments, which includes accepting funds to help set the stage for the construction of houses, provided it does not limit our ability to proclaim our Christian witness and further provided that affiliates do not become dependent on or controlled by government funds thus obtained. "Setting the stage" is interpreted to include land, houses for rehabilitation, infrastructure or streets, utilities, and administrative expenses. Funding from third parties who accept government funds with sole discretion over their use will not be considered as government funds for Habitat purposes.

Agreement to Covenant

In affirmation of the mission, method of operation, and principles stated in this covenant, we, the board of directors of the Texas Department of Criminal Justice–Hutchins Unit Chapter, covenant with Dallas Area Habitat for Humanity, Inc., other affiliates, and Habitat for Humanity International to accomplish our mission. Each partner commits to enhancing the ability to carry out this mission by supporting effective communication among affiliates, Habitat International, and regional offices; sharing annual reports; participating in regional and national training events; and participating in a biennial review and planning session between each affiliate and the regional office.

This covenant is valid upon approval by each member of the chapter board of directors and a representative of Dallas Habitat.

_____   _____
(for Dallas Area Habitat for Humanity, Inc.)   Date

_____   _____
(for the _____ Chapter)   Date

_____   _____

_____   _____

_____   _____

_____   _____

_____   _____

_____   _____

_____   _____

HABITAT FOR HUMANITY INTERNATIONAL SENIOR STAFF

Millard Fuller, President and Chief Executive Officer
David Williams, Senior Vice-President of Administration
Mike Carscaddon, Director of Finance
Regina Hopkins, General Counsel
Tom Jones, Director of HFHI Washington Office
Paul George, Director of Information Systems
Donald Patrick, Director of Human Resources
Sandra S. Byrd, Senior Vice-President of Communications and Development
Dennis Bender, Director of Communications
Roger Congdon, Director of Donor Development
Jane Emerson, Director of Corporate Programs
David Gifford, Director of International Resource Development
Alton Garrett, Director of Special Projects
Valerie Beatty, Director of Direct Marketing
Robin Shell, Senior Vice-President of Programs
Michael Willard, Director of Program Enhancement
Ted Swisher, Director of Habitat Affiliates U.S.
Ian Walkden, Director of Europe, CIS
Carrie Wagner, International Training, Education Manager
Steve Weir, Director of Asia/Pacific
Stephen Mickler, Director of Latin America/Caribbean
Harry Goodall, Director of Africa/Middle East
Alane Regualos, Deputy Director, Programs Department

Billy McGivern
Belfast, Northern Ireland

Diana Villiers Negroponte
Washington, D.C.

Nana Amba Nyarkoah Prah
Accra, Ghana, West Africa

Larry Prible
Indianapolis, Indiana

Nic Retsinas
Providence, Rhode Island

Chuck Thiemann
Cincinnati, Ohio

Sybout Vandermeer
Haarlem, Netherlands

Wayne Walker
Philadelphia, Pennsylvania

HABITAT FOR HUMANITY INTERNATIONAL OFFICES

UNITED STATES OF AMERICA
Headquarters

Director, Ted Swisher
121 Habitat Street
Americus, GA 31709
ph: 912.924.6935
fax: 912.924.6541

AFRICA–MIDDLE EAST
(Botswana, Burundi, Central African Republic, Democratic Republic of
Congo, Egypt, Ethiopia, Ghana, Kenya, Malawi, Nigeria, South Africa,
Tanzania, Uganda, Zambia, and Zimbabwe)

Director, Harry Goodall
121 Habitat Street
Americus, GA 31709
ph: 912.924.6935
fax: 912.924.0577

ASIA–PACIFIC
(Australia, Fiji, India, Indonesia, Japan, Malaysia, Nepal, New Zealand,
Pakistan, Papua New Guinea, Philippines, Republic of Korea,
Solomon Islands, Sri Lanka, Thailand, Territory of Guam)

Director, Steve Weir
Bangkok, Thailand
e-mail: asia/pacific@habitat.org

EUROPE–CIS–CANADA
(Armenia, Canada, Germany, Great Britain, Hungary, Kyrgyzstan,
Netherlands, Northern Ireland, Poland, Portugal, Romania)

Director, Ian Walkden
121 Habitat Street
Americus, GA 31709
ph: 912.924.6935
fax: 912.924.0577

LATIN AMERICA–CARIBBEAN

(Antigua and Barbuda, Argentina, Belize Bolivia, Brazil, Colombia,
Costa Rica, Dominican Republic, Ecuador, El Salvador, Guatemala,
Guyana, Haiti, Honduras, Jamaica, Mexico, Nicaragua, Paraguay,
Peru, Trinidad and Tobago)

Director, Steve Mickler
  Mailing Address:
SJO-2268
Unit C-101, 1601 NW 97th Avenue
P.O. Box 025216
Miami, FL 33102-5216

Office Address:
San Jose, Costa Rica
ph: 011.506.296.8120
fax: 011.506.232.8679
e-mail: LatinAmCar@habitat.org

Canada National Office and U.S. Regional and Area Offices

Habitat for Humanity Canada

40 Albert Street
Waterloo, Ontario N2L 3S2
CANADA
ph: 519.885.4565
fax: 519.885.5225

HFHI Washington

1511 K St. N.W.
Washington, DC 20005
ph: 202.628.9171
fax: 202.628.9169

Habitat Midwest Regional Center

(Illinois, Michigan, Wisconsin, Iowa, Minnesota,
North Dakota, South Dakota)

1920 S. Laflin
Chicago, IL 60608
ph: 312.243.6448
        800.643.7845
fax: 312.243.9632

Habitat Northeast Regional Center

(Connecticut, Maine, Massachusetts, New Hampshire, New York,
Rhode Island, Vermont, Delaware, New Jersey, Pennsylvania)

200 S. Church St.
West Chester, PA 19382
ph: 610.692.6681
     800.434.5463
fax: 610.692.7360

Habitat Southeast Regional Center

(Alabama, Georgia, Florida)

226 N. Laura St.
Jacksonville, FL 32202
ph: 904.353.1366
     800.637.9532
fax: 904.353.1544

Habitat Mideast Regional Center

(Indiana, Ohio)

11811 Shaker Blvd. #421
Cleveland, OH 44120
ph: 216.721.1200
     800.934.4684
fax: 216.721.3407

Habitat East Regional Center

(Virginia, West Virginia, District of Columbia, Maryland)

2209 Bloomsherry Dr.
Richmond, VA 23235
ph: 804.745.6156
     800.327.2530
fax: 804.674.8432

Habitat South Atlantic Regional Center

(South Carolina, North Carolina)

P. O. Box 1712
Easley, SC 29641
ph: 864.855.1102
     800.967.1102
fax: 864.850.0029

Habitat South Central Regional Center
(Tennessee, Kentucky)

P. O. Box 60410
Nashville, TN 37206
ph: 615.254.6300
      800.865.7614
fax: 615.254.8703

Habitat South Regional Center
(Alabama, Louisiana, Mississippi)

P. O. Box 4918
Jackson, MS 39296
ph: 601.355.4516
      800.283.2397
fax: 601.355.4928

Habitat Southwest Regional Center
(Texas, Oklahoma)

P. O. Box 3005
Waco, TX 76707
ph: 254.754.4673
      800.274.8177
fax: 254.756.3314

Habitat Heartland Regional Center
(Arkansas, Kansas, Missouri, Nebraska)

P. O. Box 8955
Springfield, MO 65801
ph: 417.890.6080
      800.284.0982
fax: 417.890.6090

Habitat Northwest Regional Center
(Alaska, Idaho, Montana, Oregon, Washington)

1005 N.W. Galveston
Bend, OR 97701
ph: 541.383.4637
      800.365.4637
fax: 541.383.4638

Habitat West Regional Center
(Arizona, California, Hawaii, Nevada)

1440 Broadway, Suite 205
Oakland, CA 94612
ph: 510.286.8960
    800.228.6407
fax: 510.286.8969

Habitat Rocky Mountain Regional Center
(Colorado, New Mexico, Utah, Wyoming)

1009 Grant Street #203
Denver, CO 80203
ph: 303.831.4226
    800.228.6409
fax: 303.830.1540

Acronyms Defined

AFL-CIO:	American Federation of Labor and Congress of Industrial Organizations
CNN:	Cable News Network
CEO:	chief executive officer
AREA:	Applied Real Estate Analysis
HUD:	U.S. Department of Housing and Urban Development
GED:	General Education Diploma
HVAC:	heating, ventilating, and air conditioning
CPA:	certified public accountant
CDC:	Centers for Disease Control
LSSI:	Lutheran Social Services of Illinois
NIMBY:	Not in My Backyard
HABIJAX:	Jacksonville, Florida, Habitat for Humanity
D.R.:	Dominican Republic
ACA:	American Correctional Association
ECHS:	Effingham County High School
MCC:	Mennonite Central Committee
ATM:	automated teller machine
UNC:	University of North Carolina
TDCJ:	Texas Department of Criminal Justice